REMOTE AND TELEROBOTICS

Edited by
NICOLAS MOLLET

Remote and Telerobotics,
Edited by Nicolas Mollet

Republished by InTech
Janeza Trdine 9, 51000 Rijeka, Croatia

Additional hard copies can be obtained from orders@intechopen.com

Remote and Telerobotics,
Edited by Nicolas Mollet
p. cm.
ISBN 978-953-307-081-0

For EU product safety concerns:
IN TECH d.o.o., Prolaz Marije Krucifikse Kozulić 3, 51000 Rijeka, Croatia,
info@intechopen.com or visit our website at intechopen.com.

Meet the editor

Researcher, "Workpackage Leader" at Technicolor since early 2011, I specialize in Virtual Reality (VR) and Robotics. My research currently focuses on Augmented Reality, Immersion, and Presence, to create new Technicolor applications and services. After 6 years of research and development of a training platform in VR for French military armaments (Nexter), and the participation in the assembly of a subsidiary that markets it (Nexter-Training), I managed for 3 years years a group of R & D for the realization of innovative multi-technical projects applied in the field of the (tele) Robotics, as "Team Leader" within the foundation IIT in Italy. All my work, on these 10 years of R & D, aims at real applications with high added value. My main scientific expertise focuses on human-machine interactions, on VR, communicating distributed multi-entity approaches and diffuse information. For my part, my application context led me to supervise projects with a strong technical, theoretical and human transversality. I am also a founding member of the European Association of Virtual Reality EuroVR. I was for 5 years, for the military engineering schools of Saint Cyr Coëtquidan, course leader, project manager, and head of 3 teachers. I am author of more than 25 international publications (patents, book and book chapters, newspapers, conferences).

Preface

Any book which presents works about controlling distant robotics entities, namely the field of telerobotics, will propose advanced technics concerning time delay compensation, error handling, autonomous systems, secured and complex distant manipulations, etc. So does this new book, *Remote and Telerobotics*, which presents such state-of-the-art advanced solutions, allowing for instance to develop an open low-cost Robotics platform or to use very efficient prediction models to compensate latency. This edition is organized around eleven high-level chapters, presenting international research works coming from Japan, Korea, France, Italy, Spain, Greece and Netherlands.

The particularity of this book is, besides all of those innovative solutions, to highlight one of the fundamental tendency that we can see emerging from this domain, and from the domain of Human-Machine interactions in general. It's a deep reflection, aiming to redefine this problematic of interaction spaces divergence: a human acts according to his own models of perception-decisionaction, fundamentally different from the machine's ones. Those models cannot be identical by nature, and rather than transforming the human into an expert *adapted* to a very particular task and its according dedicated interface, those deep reflections try to characterize precisely the way to transform one interactions space to another. Thus the second moiety of the book regroups a set of works which integrate those reflections. It concerns for instance the identification of objective characteristics and parameters dimensioning the human in this context, to take into account his own evolution, or also to design interfaces that he can natively identify and use thanks to a natural empathy and appropriation.

Despite a constant technological development, always more specific, surprising and innovative, several domains like the *teleoperation* one have identified obstacles to some important conceptual and technological innovations, which have prevented for example the ambitious engagements of personal robots fully autonomous and intelligent by the end of the previous century. While the human accommodates frequently himself to the technologies he creates, he becomes now one of the main limitation of some important technological breaks, because of his own ignorance about himself. At the time of technologies which have deeply transform our life and our future, sometimes in dangerous ways, those reflections allow us in the meantime to think about human's particularities, evolutions and needs. To go steps forward, the human needs to better understand himself: we found here one of the fundamental and natural goal of science, namely to understand, to know, to better determine what and who we are, where we come from and where we are going to.

Nicolas Mollet
03/23/2010
TEleRobotics and Applications (TERA) Dept.
Italian Institute of Technology (IIT)

Contents

1

Electronics proposal for telerobotics operation of P3-DX units

Felipe Espinosa, Marcelo Salazar, Daniel Pizarro and Fernando Valdés
University of Alcala. Electronics Department
Spain

1. Introduction

Telerobotics is the area of robotics concerned with the control of robots from a distance, mainly using wireless connections or the Internet. It is a combination of two major subfields, teleoperation and telepresence. The work presented in this chapter belongs to the field of teleoperated robots, where a remote centre sets commands to the robot and supervises the performed motion by receiving feedback from its sensors. In teleoperated robots the control algorithm can be balanced between the remote host and the local host in the robot, which yields to several kind of possible control schemas.

The key components needed to develop telerobotics applications are the following: control (algorithm and real time implementation), sensors (world sensing and information processing) and wireless communication (generally using standard wireless technologies, i.e. IEEE 802.11) [Angelo, 2003], [Anvari, 2007], [Gumaste, 2007], [Mehani, 2007], [Chumsamutr, 2003], [Hespanha, 2007], [Bambang, 2007].

This chapter is outlined within both educational and research fields in telerobotics, and so its aim is to offer a reliable and low cost architecture to be implemented in research labs. The robotic platform consists of the Pioneer 3DX (P3-DX) from the company MobileRobots (see Figure 1). It is made of an aluminium body (44x38x22cm) with 16.5cm diameter drive wheels. The two DC motors use 38.3:1 gear ratios and contain 500-tick encoders. The differential drive platform is highly holonomic and can rotate in place moving both wheels, or it can swing around a stationery wheel in a circle of 32cm radius. A rear caster is included for balancing the robot. On flat floor, the P3-DX can move at speeds of 1.6 mps. At slower speeds it can carry payloads up to 23 kg. In addition to motor encoders, the P3DX base includes eight ultrasonic transducer (range-finding sonar) sensors arranged to provide 180-degree forward coverage. This robot includes a 32-bit RISC-based controller, with 8 digital inputs and 8 digital outputs plus 1 dedicated A/D port; 4 of the outputs can be reconfigured to PWM outputs [P3-DX, 2009].

The P3-DX can be ordered with a complete electronic hardware [MobileRobots, 2009], which include wide range sensors, an on-board PC and Wireless Ethernet communication device. However, the authors propose to start from a basic structure that allows to be customized depending on the final application. This decision offers the opportunity of working with open platforms which is specially suitable for educational labs in engineering schools. On the other hand, the final cost of the prototypes is substantially reduced using general purpose hardware and developing ad-hoc software as it is detailed next.

Fig. 1. Basic robotic platform of Pioneer 3-DX.

In the context of telerobotics some questions must be addressed: which are the features that a robot must have to be teleoperated and how to provide a robotic platform with low cost devices so that such features are implemented.

To become a teleoperated robot, three subsystems are needed: control, communications and sensors.

From the control side three levels are proposed in this document:

- Low level control (LLC), which directly controls the active wheels of the robot. The P3-DX includes a PID for each active wheel [P3-DX, 2009].
- Medium level control (MLC), for path following. In this document a linear servo system is proposed. [Ogata, 1994], [Dutton et al., 1997].
- High level control (HLC), where a more complex control is required and extra sensors which give richer information about the environment. As an example, in platooning applications, the HLC determines the path by sensing the distance and relative position of the preceding follower, [Kato et al., 2002], [Espinosa et al., 2006].

From the communications side, a wireless network (short range for indoor applications) is required with a topology depending on the application: Using a star network topology, one or several teleoperated robots behave as wireless nodes whose master node is the remote centre, (see Figure 2.left). In applications where all robots must share the same information a fully connected mesh topology is preferable (see Fig. 2.right).

From the point of view of the sensor included in the robot, both the application and the environment drive the quality specifications and amount of information required for following the commands sent by the remote centre.

If the environment is free from obstacles and the paths are not very large, the odometry information included in the P3-DX can be an option. However, if the paths are large or repetitive, the accumulative error present in the dead-reckoning techniques must be compensated with an absolute localization method (e.g. vision sensors or infrared-beacons)

[Borenstein et al., 1996]. If the application requires the detection of obstacles with a field of view of 180°, 5 meters of depth and a 0.1 feet resolution, the built-in sonar system in the P3-DX platform can be reliable and enough. Contrary, if more accuracy is needed a laser-range sensor, such as the Hokuyo Scanning Laser Range Finder [Hokuyo, 2009] is proposed.

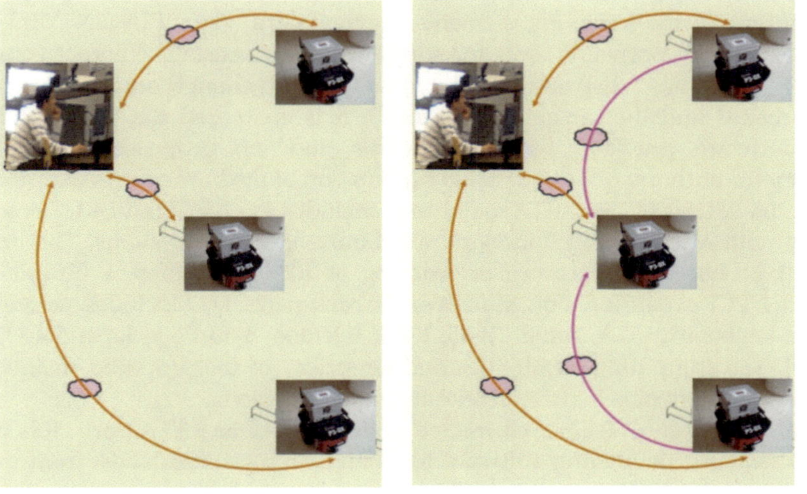

Fig. 2. Example of telerobotics operation: without (left) and with (right) cooperation among robot units.

The basic hardware included in the P3-DX is not enough for supporting the control, communication and sensing requirements in robotic teleoperation. According to the authors' experience, the minimum specifications are the following:

- Embedded PC with native x86 architecture and at least: 2 USB ports, 1 firewire header, on-board LAN, 1 SATA connector, and mouse and keyboard ports.
- SATA Hard Disk of 10 Gb, to save long experimental data.
- Wireless ethernet converter, server and client modules, allowing security system and transmission rate superior to 10 Mbps.
- Additional sensor system to improve the obstacle detection.
- Real Time Operating Systems for control and communications tasks implementation.
- Development tools for low level robotics applications.

2. Hardware architecture

In the previous section, the basic hardware of P3-DX has been presented as it is shipped from MobileRobots in its basic configuration (See Figure 1). The more relevant subsystems of this electronic architecture are: Hitachi microcontroller, encoders, sonar ring and the global power electronics from a battery pack. The microcontroller is in charge of, among other functions, executing the LLC loop (PID) of each motor in the active wheels (Left and Right). The LLC obtains feedback from the odometry sensors in the wheels [P3-DX, 2009]. Graphically, the block diagram of this electronic architecture is showed in the Figure 3. (left part).

In the following lines the hardware and software, designed specifically for the teleoperation of the P3-DX robot are described.

2.1 Hardware components

Taking into account the aforementioned specifications, the hardware proposed in this chapter consists of the following elements: motherboard VIA EPIA EN Mini-ITX, which incorporates enough ports and slots for supporting new sensors. Ethernet converter WLI-TX4-G54HP, 80 Gb hard disk and DC-DC converter with enough output power for driving the motherboard and the sensors. In Figure 3, it is compared the block diagram of the original hardware provided by MobileRobots and the proposed hardware add-ons proposed by the authors. A more detailed explanation of the hardware is described next.

The VIA EPIA EN15000G Mini-ITX mainboard includes the 1.5 GHz VIA C7 processor, fully compatible with Microsoft® Windows® and Linux operating Systems. This motherboard integrates 1 on-board LAN controller working at 100/1000 Mbps, 1 IEEE 1394 firewire header and 1 PCI expansion slot. Moreover its back panel I/O includes several ports: PS2 mouse and keyboard, VGA, serial, RJ-45, RCA, S-Video, 3 audio jacks and 4 USB 2.0 [Via-Epia, 2009]. The computing capabilities and versatility of the proposed mainboard allows the easy development of robotics and telerobotics applications.

An external hard disk of 80 Gb is connected to one of the free SATA ports, this way enough capacity is kept for debugging software tools and saving information from experimental tests.

The different voltage levels required by the hardware (i.e. +3.3V, +5V y +12V) are provided by the MOREX QT-16045 DC-DC converter.

In order to implement the wireless communication, required in a telerobotics application, see Figure 2, two modules are needed: a router (generally in the remote centre) and a client for each teleoperated robot available in the network. The wireless router chosen for this application is the Buffalo WHR-HP-G54, which is in compliance with standards IEEE802.11b/.11g, offers 11 frequency channels and allows high-speed transmission rate of 125 Mbps. Others characteristics are: wireless security WPA-PSK and 128 bits WPE, 4 LAN ports, high-speed routing. The client chosen to connect to the router is the WLI-TX4-G54HP [Buffalo, 2009].

In Figure 4 it is shown the appearance of the electronic hardware detailed before, whose cost (September 2009) is under 700 $ USA.

Some details about the P3-DX platform are shown in Figure 5, incorporating the proposed electronics subsystem. The prototype of the picture is part of the research developed by the authors in the COVE project (Intelligent Transport System for Cooperative Guidance of Electrical Vehicles in Special Environments) held in the Electronics Department at the University of Alcala (Spain).

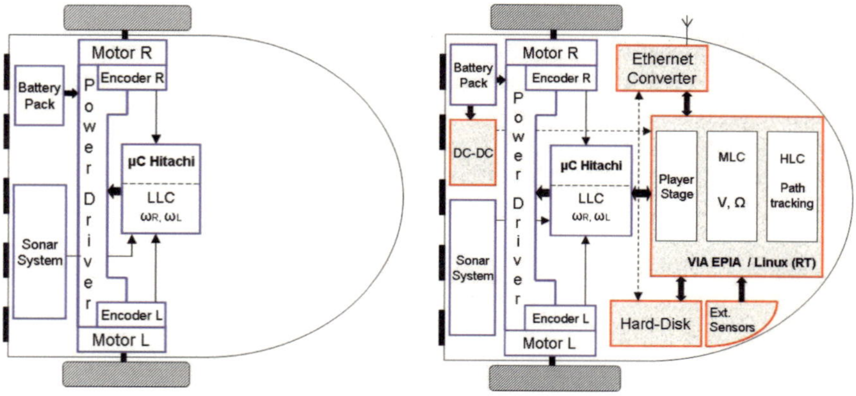

Fig. 3. Block diagram of the P3-DX basic electronic architecture and the add-ons proposed for telerobotics applications.

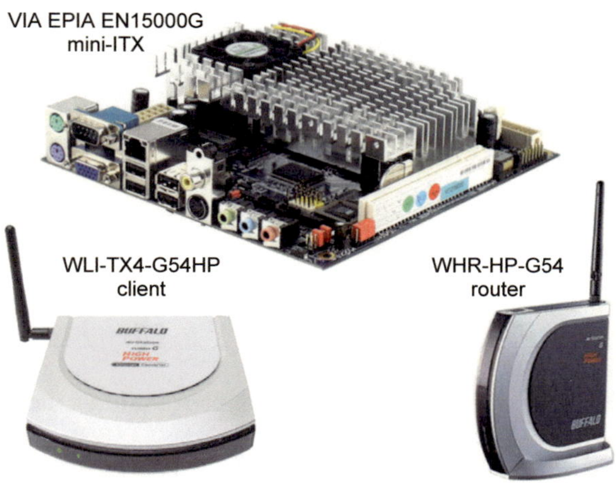

Fig. 4. Main electronics devices involved in the designed architecture for teleoperation of the P3-DX.

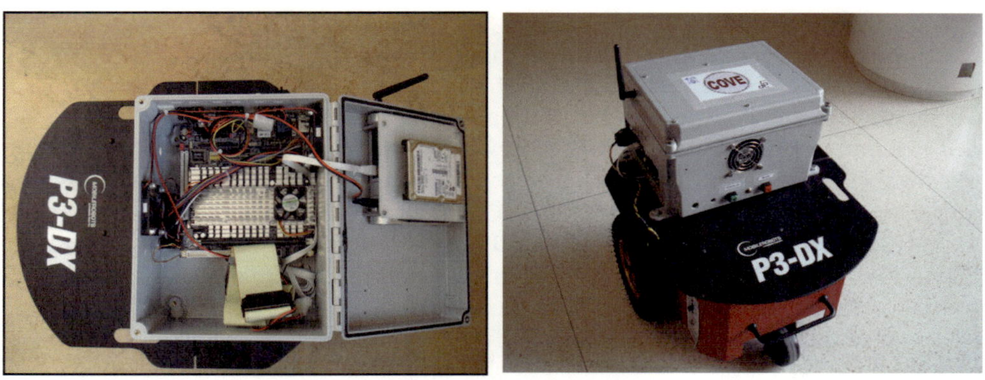

Fig. 5. Photographs of the P3-DX modified version implementing the new hardware architecture hardware for telerrobotics applications.

2.2 Software components

In order to develop the teleoperation capabilities aforementioned, a multilevel software platform is proposed, which combines high level UML design and control, given by the platform Matlab/SIMULINK [MathWorks, 2009], and the low level client-server software PlayerStage [PlayerStage, 2009], which communicate with P3DX microcontrolers and all sensors in the robot. Both client/server platforms are hosted in Linux/RTAI [RTAI, 2009] soft real time operating system to ensure accurate control over periodic tasks and low latency.
The general diagram of the software platform as well as the input/output topology is shown in Figure 6.

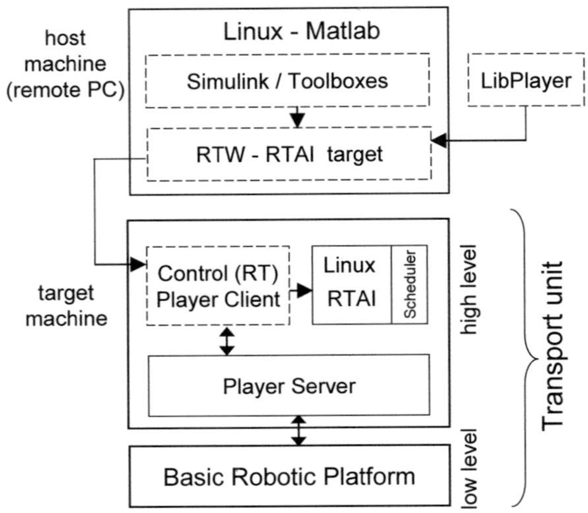

Fig. 6. Software tools integration for control design in telerobotic applications.

Roughly speaking the software architecture can be described in two main parts, the client-server architecture based on Player/Stage and the Matlab/Simulink driver using the Real Time Workshop (RTW) framework [MathWorks, 2009].

2.2.1 Client-server architecture based on Player/Stage

The Software Player/Stage (P&S) consists of a complete suite for control, communication and simulation of mobile robots under UNIX operating systems [PlayerStage, 2009].
Player is a network interface for robotic devices: controllers, sensors and actuators. It is a hardware abstraction layer that allows the coordination and distribution of tasks within and between robots. The client-server model of Player easily allows robot control programs to be executed locally or remotely in a connected remote centre. In addition, the open structure of Player allows writing the algorithms in diverse programming languages and executing them in Real Time operating systems (i.e. Linux RTAI).
For the case of the P3-DX platform, Player offers a standard interface for mobile robots that allows sending high-level commands (angular and linear velocity) for motion in a 2D surface as well as dead-reckoning functions for retrieving the 2D position of the mobile

robot [Whitbrook, 2006]. In addition, player offers direct access to sensors onboard the robot, such as laser range or ultrasound rings.
Stage is a simulator of multiple mobile platforms moving in a plane (2D). More precisely, it provides a virtual world for robots and objects that can be detected by sensor systems on-board the robots. This software makes easier the transition between simulation and the physical implementation of the control algorithms.
In order to perform movements with the robotic platform, it is necessary to run at least two programs in coexistence: the P&S server and the client one (see Figure 7).

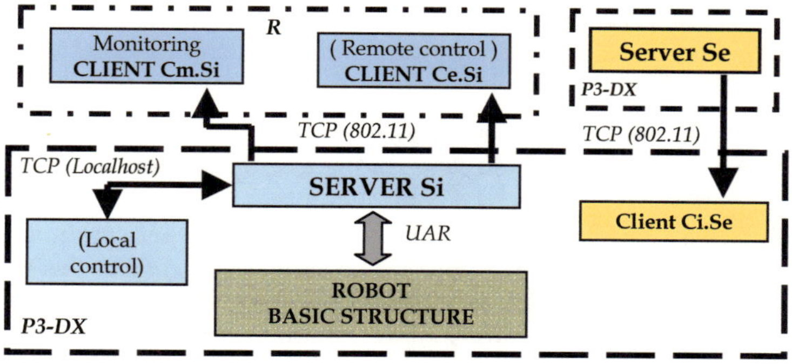

Fig. 7. Proposal of client-server architecture based on Player/Stage for the development of telerrobotics application with P3-DX units.

The server (Player Server), namely S_i , establishes communication with the robot hardware (in this case the basic structure of the P3-DX) using a serial link RS-232 (UART). The server offers information about the state of the robot and accepts orders to control it.
Connected to the S_i Player Server, there can be clients that are executed either locally (e.g. $C_i.S_i$) or externally to the robot hardware (e.g. $C_e.S_i$ and $C_m.S_i$). The control program, which is always installed in client mode (Player Client) $C_1.S_i$, at each sample time T_s analyzes the state of the robot and decides the movements to be performed. In general there is only one control program for each robot whether it is locally executed $C_1.S_i=C_i.S_i$ or run in a remote centre (RC) $C_1.S_i = C_e.S_i$.
In Figure 7, it can be shown two possible clients of the server installed in the P3-DX robot:

- Player client $C_m.S_i$, is a monitoring program installed for example to detect the location (position and orientation) or changes in motion (linear and angular velocity) of the robot.
- Player client $C_e.S_i$, performs several different tasks, as for example the control of the robot from the remote centre. In this configuration the RC is in charge of both monitoring and control.

The proposed client-server architecture allows a robot to execute clients which connect to other servers, namely $C_i.S_e$ in Figure 7. This functionality is of great interest in cooperative guiding and robot platoon applications [Espinosa et al., 2008], where the controller of one unit needs to monitor the state of some other robots. To sum up, each server S_i, executed in a robotic unit can receive connections from the following client processes:

- Client embedded in the robotic unit $C_i.S_i$, with the tasks described for $C_1.S_i$ if the control is local.

- Monitoring client $C_m.S_i$ implemented in the RC.
- Clients embedded in other transport units $C_e.S_i$, with tasks of $C_1.S_i$ if the robot control is remote mode.

The physical medium that supports the local client-server communication is transparent to the Player/Stage user. In remote mode this communication can be implemented using TCP/IP protocols over Ethernet (IEEE 802.3) or Wireless (IEEE 802.11). On the other hand, when the client is hosted in the same CPU as the server (e.g. local control configurations), the client-server communication keeps the TCP/IP layer over a loopback local device. This feature allows modifying control configuration with very little modifications in the design. This transparency is shown in Figure 7, were local ($C_i.S_i$) and remote clients ($C_e.S_i$) connect to the same Player server.

2.2.2 Matlab/Simulink driver in Real Time Workshop

The Matlab software offers a complete numerical/symbolical engine for scientific simulation. It is widely used among the control community, specially its integrated UML module, called Simulink, designed for simulation and control over continuous and discrete dynamical systems. Thanks to the Simulink environment, the control design is performed and tested in the same language, i.e. using differential equations and transfer functions to describe controllers and models. Simulink allows communicating with real peripherals, such as serial ports, network sockets or any other kind, generating custom blocks or "S-functions" which are integrated in the design. Besides simulation, the combination of Simulink with the Real Time Workshop (RTW) [MathWorks, 2009] tool allows compilation and execution of the designed controller in different targets, such as FPGAs, Digital Signal Processors, Microcontrollers or PCs with real time operating systems as it is the case in this paper.

By using custom blocks generated for Simulink it is possible to generate an S-function block that connects with a P3-DX platform which is running a Player server through a TCP/IP socket. The position of the robot and the state of all its sensors can be sent to the Simulink controller where a block-diagram like algorithm can be easily designed.

In this document, several control configurations are proposed, whereas the control algorithm is executed locally in the robot or externally, using always the framework offered by Simulink and the RTW compiling for Linux RTAI targets. In Figure 8 it is shown the common driver used to perform control of a P3-DX platform using the aforementioned Simulink/RTW schema. The block is composed of two main parts:

- The first part consists of an S-function or father process which is executed at each time sample driven by the inputs from a signal builder in the Simulink environment. The father process communicates using shared memory buffer with a child process that is in charge of network communications.
- The child process, which is executed in parallel, implements socket communication with the Player Server hosted in the robot. It sends angular and linear speed commands and receives position and sensor readings, which are sent back to the father process. The main justification of this father-child methodology is to separate blocking connections in the TCP/IP socket from the S-function main loop.

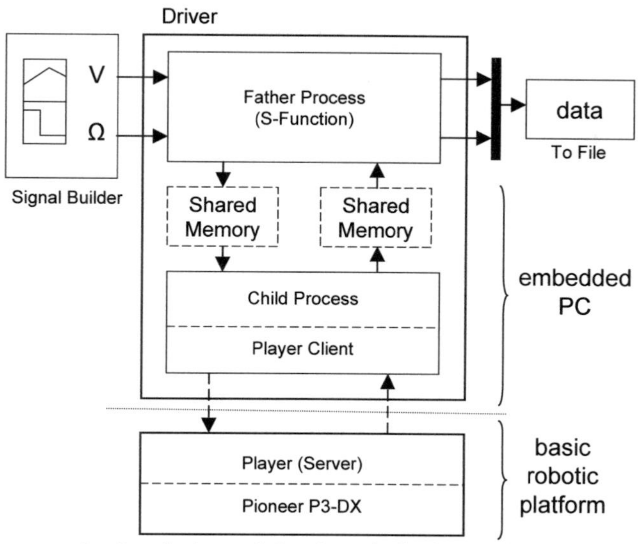

Fig. 8. Involved process in the design of the Simulink-driver for the low level access of the P3-DX platform.

Using the driver presented before, several control schemes are defined, which can be executed locally and remotely by only changing few settings in the target obtained by RTW.

3. Teleoperation control schemes

In this section, several teleoperation control schemes are described with the aim of showing the capabilities of the proposed architecture for such kind of applications. Therefore, a simple controller is suggested for tracking the linear and angular speeds of the robot, which was mentioned in the Introduction section as the middle level control (MLC). Then, two modes of operation are presented, where the controller is hosted externally to the robot's hardware (remote servocontroller) and directly executed inside (local servocontroller). Using the stated software/hardware architecture, the different teleoperation modes only require minor design modifications.

3.1. Middle level control of the robot

The independent LLC for each wheel (PID) is in practice not enough to correctly control the angular and linear speeds of the robotic platform. The authors propose to use a servosystem controller as the Middle Level Controller that guarantees null error in steady state for constant references and low constant error for ramp references of angular and linear speed.
First, a parametric identification of the robot is carried out by means of a linear and time invariant state-variable model G, H, C, D, where the state variables are related to both wheels speeds at different sample times, and the inputs are obtained from the desired linear and angular commands sent to the robot. The state-variable model is dependent of the LLC and eventually can replace it if the local PIDs of the wheels are disconnected.

As it is shown in Figure 9 the servosystem is designed to follow angular and linear speed commands in the robot. The controller has several degrees of freedom in its design, affecting its performance. The gains K_r and K_i are set using a Linear Quadratic Regulator (LQR) [Ogata, 1994], [Dutton, 1997] approach.
As was commented before, the P&S architecture makes possible the implementation of the MLC in two versions (local and remote) for telerobotics applications. In both cases, the motion reference or motion set-points are imposed by the remote centre.

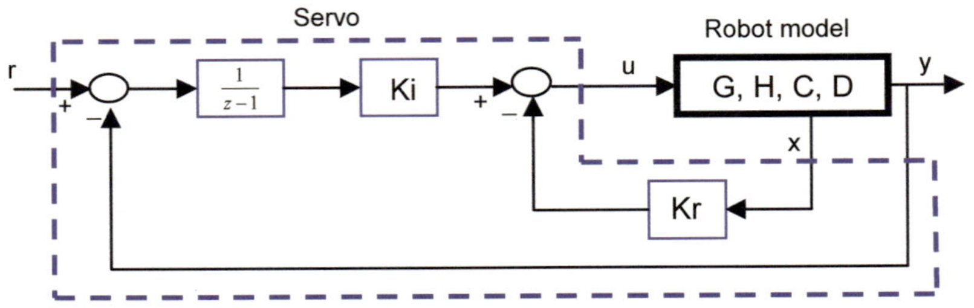

Fig. 9. Block diagram of the designed servosystem for robot velocities tracking.

3.2. Telerobotics application of P3-DX through remote servocontroller

In this alternative, the middle control level tasks (global velocities tracking of the mobile unit) are carried out in an external computer that serves as remote controller, see Figure 2. The commands given to the LLC to follow a given trajectory are established in the RC, the instantaneous error is calculated from the information received by the robot on-board sensors using the wireless communication channel. In the same way the resulting control outputs are sent to the mobile platform using the wireless channel. This control strategy is compatible with the Player/Stage architecture: the control algorithm in the RC is defined as a client $C_{RC}.S_R$ of the server S_R in the P3-DX robot. A general diagram of the elements involved in this solution for control and communication can be seen in Figure 10.

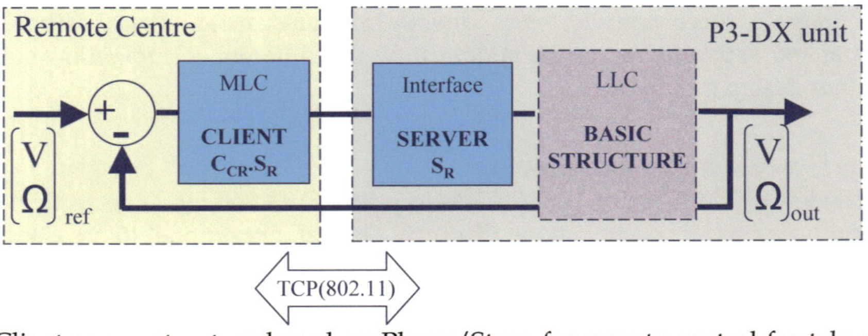

Fig. 10. Client-server structure based on Player/Stage for remote control for tele-operation of P3-DX units.

This is a clear example of how an external client links to the on-board server, no matter if they are implemented on different hardware platforms. Player/Stage satisfactorily handles these kinds of information transfer.
The remote servocontroller allows to minimize the hardware required in the robotic unit, however the system stability can be compromised due to well known problems derived from the wireless channel (delays, packet dropout, limited bandwidth,...).

3.3. Telerobotics application of P3-DX through local servocontroller

This alternative considers the implementation of both control levels LLC and MLC onboard the robot, see Figure 11.
The RC is only in charge of sending the commands for the desired movement. This task requires a non periodical updating time that generally is higher than the control sample time T_s. This is why a complementary client-server service based on sockets is implemented making easier the P3-DX teleoperation. In this way the client $C_R.SS_{RC}$ in the robot unit is updated with the reference locations from the server SS_{RC} for local control set-points.
The primary advantage of this proposal is that the critical control tasks are executed in the robot client $C_1.S_R$, which means higher immunity to wireless communication channel. On the other hand, the $C_R.SS_{RC}$ update, which is directly affected by the channel, can be asynchronous and longer.

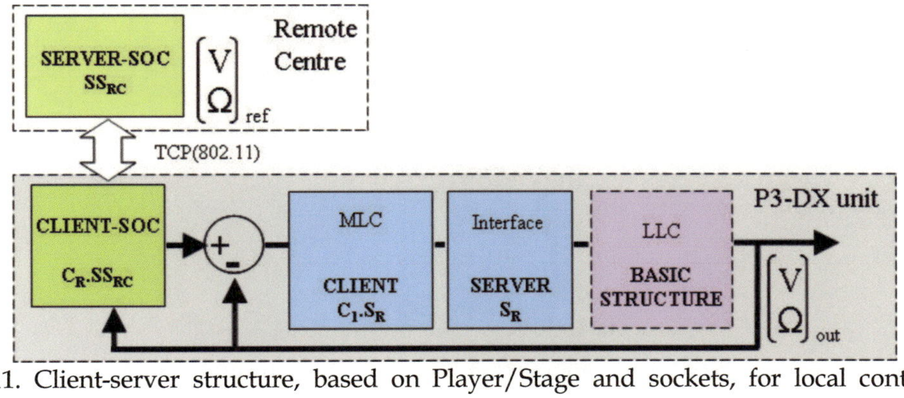

Fig. 11. Client-server structure, based on Player/Stage and sockets, for local control of teleoperated P3-DX.

4. Experimental results

Both hardware and software architectures described in this chapter are being used in different robotic applications inside the aforementioned COVE research project. In this section some experimental results are presented which support the pros and cons of the proposed architecture.
In Figure 12, a typical laboratory work session is shown, in which the human operator uses a PC as a remote center of the teleoperated P3-DX robot, which includes the designed hardware and software add-ons. As an example of the system capabilities, a path involving changes in angular and linear speeds is sent to a single robot. The teleoperated unit is

running the servo controller designed in section 3.1, and the sensor systems consists of odometry readings.
In this experiment the communication protocol consists of the UDP protocol (instead of TCP) allowing thus packet loss between the client and server. In the case that a command is lost, it is replaced with the one from previous instant. Such way of managing the packet loss is done in two directions: RC-robot (command sent), and robot-RC (odometry readings).
In a first experiment, the speed control for path following is implemented in local mode as it is stated in section 3.3. In this case, the remote center is only in charge of sending commands and supervising the result, but the MLC is hosted in the robot's CPU.
A second experiment is proposed, whose objective is to track the same commands as in the local mode, but the MLC is implemented in the remote center, following the idea presented in section 3.2. Therefore, only the LLC (PID) is executed in the robot, whilst the command generation, control of angular and linear speeds (MLC) and path monitoring are executed in the remote center.

Fig. 12. View of the laboratory set-up of telerobotics applications with P3-DX units.

In Figure 13 it is shown a temporal evolution of the linear speeds in the following cases:

- The linear speed performed using the local mode is shown in blue. As it can be observed, the speed (m/s) is following a trapezoid shape as it is expected in the experiment.
- The linear speed performed using the control in remote mode, and supposing that there is no packet loss is displayed in red. In the inferior part of the figure, where a zoom region of interest is presented, the channel communications includes two delays (two zero samples at the beginning), one for each direction in the channel.
- In yellow it is presented the linear speed of the robot in remote mode but including some percentage of packet loss and keeping, as was commented before, the previous command in that case. This experiment is performed with an approximate rate of 5% of packet dropout.

- The same experiment than the previous one but supposing a packet loss percentage around 50% is displayed in black. In this case it is expected that the trajectory performed by the robot derives away from the desired one.

In order to stress the risks of including a wireless cannel inside a teleoperation process in remote mode, Figures 14 and 15 are shown. Figure 14 shows six realizations of the same trajectory followed by the robot in different control configurations. The three trajectories displayed in green belong to the MLC in local mode and, as it is observed, the differences between the three are negligible. The three paths displayed in red consists of the MLC in remote mode assuming a 5% of packet dropout, which as it is expected increase the differences between the paths.

The Figure 15 shows the same experiment but including a 50 % of packet dropout. In this case the channel is highly corrupted, which can invalidate the teleoperation capabilities, especially in the remote mode. The main problem of this level of packet loss is not that the target is not reached exactly as it is expected, but that the effects on the trajectory are highly random.

It must be remarked that the experiment has been performed using only the odometry readings from the robot. If more sensors are included, which allow an absolute positioning device (e.g. sonar, laser, vision…), the deviations from the desired path can be minimized.

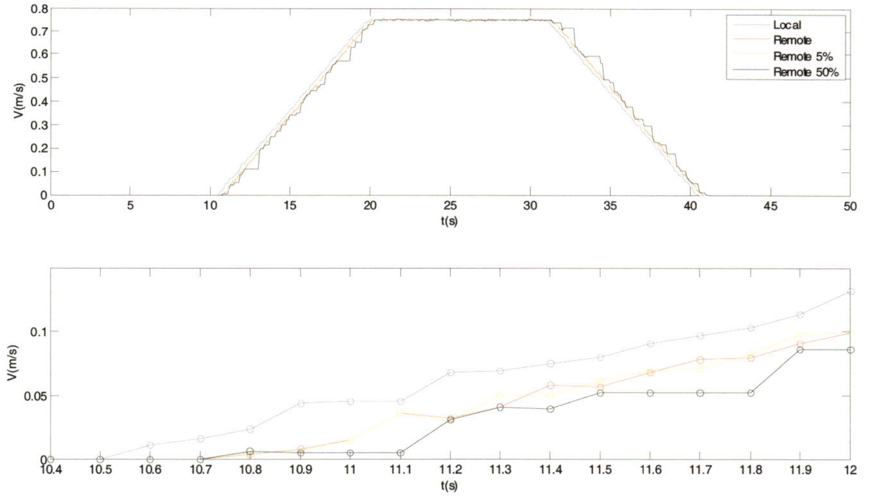

Fig. 13. Comparison of linear velocities from different tests of local control and remote control (UDP with packet dropout).

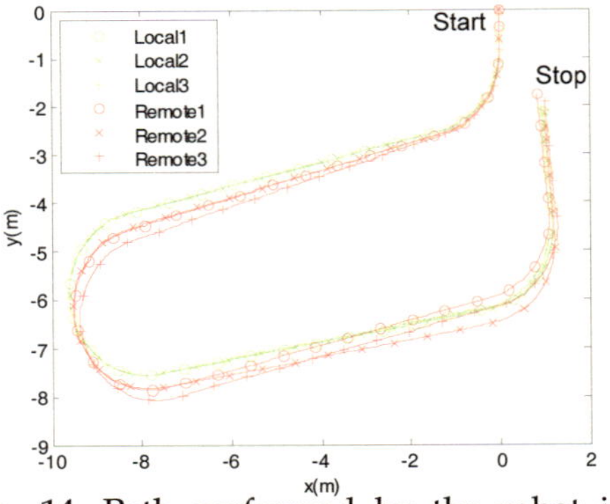

Fig. 14. Path performed by the robot in both local and remote mode (UDP with 5% packet dropout).

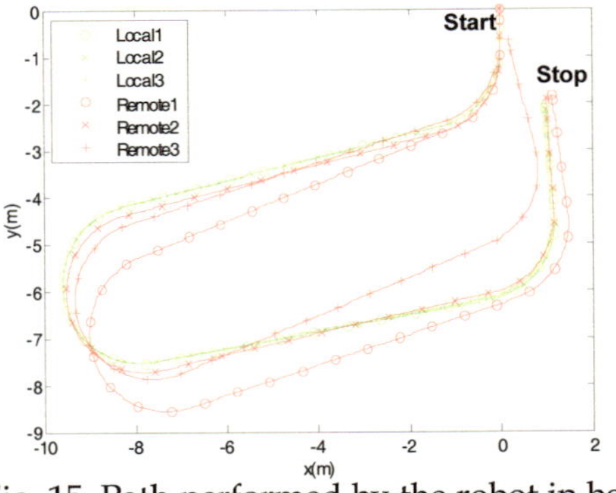

Fig. 15. Path performed by the robot in both local and remote mode (UDP with 5% packet dropout).

5. Conclusion

In this chapter the authors describe a solution for providing a basic commercial robot with the capabilities of being teleoperated. The proposed electronics architecture fits the requirements in terms of sensor processing, control and wireless communications needed in a telerobotic application.
From the hardware point of view, the key elements are the motherboard Via-Epia miniITX, and the wireless hardware included in the remote client WHR-HP-G54 and the one mounted on the robot WLI-TX4-G54HP.
From the software side, the more relevant feature is the high flexibility which is provided from the synergy between Matlab/Simulink/RTW with the client-server structure of the Player/Stage open software for the implementation of control applications.
Together with theoretical descriptions, examples of real telerobotics applications are shown. For that, the ad-hoc configured robotic platform is tested, working in both control modes: local and remote. Properties and drawbacks of using a wireless channel inside a control loop are remarked.
In practice, the prototype carried out by authors is being used as a real test bench for telerobotics and control applications, including robot cooperation problems in both educational and research fields.

6. References

Angelo, J.A. *Robotics: a reference guide to the new technology.* IEEE Communications Magazine. Volume: 41 , Issue: 12, page(s): 60 67. Dec. 2003 ISSN: 0163-6804. INSPEC Accession Number: 7950580

Anvari, M. *Telementoring and remote telepresence surgery. Chapter of the book "Robotics in Surgery: history, current and future applications".* ISBN: 1-60021-386-3. Nova Science Publisher, Inc 2007

Bambang, R. *Development of architectures for internet telerobotics systems.* Proceedings of the International Conference on Intelligent Unmanned System (ICIUS 2007), Bali, Indonesia, October 24-25, 2007, Paper No. ICIUS2007-B004

Borenstein, J.; Everett, H.R. ; Feng, L. (1996). *Where am I?. Systems and Methods for Mobile Robot Positioning.* Edited and compiled by J. Borenstein. March 1996. Electronic copies of this report in its entirety can be downloaded from http://www-personal.umich.edu/~johannb/shared/pos96rep.pdf

Buffalo (2009). WLI-TX4-G54HP Datasheet available: http://www.buffalotech.com/files/downloads/WLI-TX4-G54HP_DS.pdf

Chumsamutr, R. and Fujioka, T. *Development of Car-Following Model with Parameters Identified by Genetic Algorithm.* JSME Int Journal. Ser C. Mech Systems, Mach Elem Manuf. Journal Code: X0995A. ISSN: 1344-7653. VOL.46;NO.1;PAGE.188-196 (2003)

Dutton, K; Thompson, S.; Barraclough, B. (1997). *The art of control engineering.* Addison-Wesley, 1997. ISBN- 0-201-17545-2

Espinosa, F.; Awawdeh, A.M.H. (2006). *Focus on Robotics Research Book. Chapter: New strategies and control algorithms to reduce platoon oscillations in linear as well as non-linear trajectory.*Editor: John X. Liu , Nova Science Publishers, Inc. Hauppauge, New York. 2006

Espinosa, F.; Salazar, M.; Bocos, A.; Valdés, F.; Iglesias, R. (2008). *Design and Implementation of a Communication Architecture based on Player/Stage for Telerobotics Operation of P3-DX units.* International Conference on Robotics and Automation -ICRA 2008-. Workshop: New Vistas and Challenges in Telerobotics. IEEE Catalog Number: CFP08RAA-CDR, ISBN: 978-1-4244-1647-9, ISSN: 1050-4729. Pages: 65-70. Pasadena, California, USA. May 19-23, 2008

Gumaste, A.; Singhai, R. and Sahoo, A. *Intelligent Vehicular Transportation System (InVeTraS).* Telecommunication Networks and Applications Conference, 2007. ATNAC 2007. Australasian. 2-5 Dec. 2007 Page(s):58 – 63. Digital Object Identifier 10.1109/ ATNAC.2007.4665283

Hespanha, J.P.; Naghshtabrizi, P. and Xu, Y. *A survey of recent results in Networked Control Systems.* Proceedings of the IEEE, vol. 95, no 1, pp. 138-162, January 2007

Hokuyo (2009). http://www.active-robots.com/products/sensors/hokuyo.shtml

Kato, S.; Tsugawa, S.; Tokuda, K.; Matsui, T. and Fujii, H. (2002). *Vehicle Control Algorithm for Cooperative Driving with Automated Vehicles and Intervehicle Communications.* IEEE Trans. on Intelligent Transportation Systems, Vol. 3 , nº 3, pp 155-161

MathWorks (2009). http://www.mathworks.com/products/rtw/

Mehani, O.; Benenson, R.; Lemaignan, S. and Ernst, T. *Networking needs and solutions for road vehicles at Irnara.* ITST '07. 7th International Conference on ITS . ISBN: 1-4244-1178-5. Digital Object Identi_er: 10.1109/ITST.2007.4295894

MobileRobots (2009). http://www.mobilerobots.com/

Ogata, K (1994). *Discrete-time control systems.* Prentice Hall, 2 edition. December, 1994. ISBN-10: 0130342815

P3-DX (2009). http://www.activrobots.com/ROBOTS/p2dx.html

PlayerStage (2009). The Player Project: http://playerstage.sourceforge.net

RTAI (2009). RTAI-Linux Target HowTo. http://www.mathworks.com/matlabcentral/files/10742/RTAI-TARGET-HOWTO.txt

Via-Epia mini-ITX (2009). Datasheet available: http://www.via.com.tw/en/products/embedded/ProductSeries.jsp?serialNo=2
Whitbrook, A. *Controlling the Pioneer P3-DX robots at CSiT*. University of Nottingham. April 2006

2

Decorators Help Teleoperations

Shinichi Hamasaki and Takahiro Yakoh
Keio University
Japan

1. Introduction

The wide bandwidths of the communication and advanced bilateral robot control technologies have led us to realize multi-sensational teleoperation systems that provide auditory, visual, and haptic information of remote site to its operator simultaneously. Such teleoperation system is expected to be applied to medicine, education, work in hazardous environment, online game or other use. Especially, haptic sensation is extremely required to improve its operability and safety in medical operations. In this chapter, a teleoperation system is assumed to consist of three components, i.e., audio, video, and haptics transmission systems. In addition, each transmission system can be distinguished into local site and remote site.

When the distance between the remote site and the local site of a teleoperation system is far enough, the performance of its communication line becomes crucial factor to decide the operability of the teleoperation. In general, since the data rate of video information is higher than the bandwidth of communication line, data compression is indispensable. Moreover, data compression and decompression are considered as time consuming processes. As a result, video transmission system must be delayed in principle. In multi-media context, audio transmission system should be delayed artificially so as to keep its playout delay to be the same as that of video transmission system. For example, lip sync is necessary for its operator to feel the audio and video contents in naturally. From this point of view, the bandwidth of its communication line is the only requirement for realize video and audio transmission systems. Even if these transmission systems are used to make a conversation for its operator with a remote person, it is said that 200ms delay is allowable. This requirement is rather negligible in the Internet in nowadays. On the other hand, a haptics transmission system requires short and stable performance of delay for communication line. This is because the achievable bandwidth of haptic sensation is limited by the round trip time of communication line since haptics transmission controller includes the delay inside of its closed control loop. In fact, a human being can feel the sense of touch at the tip of a finger up to about 400Hz. To recognize this bandwidth of haptic sensation, its sampling period must be higher than 800Hz according to Shannon's sampling theory. Thus, the round trip time of communication line must be shorter than 1ms. This value is much shorter than the allowable delay of audio and video transmission systems. In short, haptics transmission system requires short delay while video transmission system requires wide bandwidth for communication line.

If haptics transmission system and video transmission system are constructed individually, these systems will not synchronize at all. The operator may hope these two systems synchronize although the operator may feel strangeness with such unsynchronized teleoperation system. If the haptics transmission system is delayed artificially in order to synchronize video transmission system, the bandwidth of haptic sensation will also be limited accordingly and the overall system's operation will deteriorate. Likewise, if the delay of video transmission system occurs to synchronize haptics transmission system, the quality of video, such as, resolution, depth of color, and frame rate, will be reduced drastically. Therefore, it is impossible to realize both of high quality of each service and synchronization at the same time with infinite bandwidth communication line. This is the underlying issue of this chapter.

This chapter proposes decorators to lessen the strangeness of such unsynchronized teleoperation system. The latter part of this chapter is organized as follows: Section 2 overviews related works to this issue. Section 3 includes our proposed decorators. A target task designed for operability evaluation of teleoperation and its experimental equipments are shown in Section 4. Section 5 explains the implementation of proposed decorators. Experimental results are shown in Section 6. Finally this chapter is concluded in Section 7.

2. Related Works

Many researchers and developers have been trying to improve the operability of teleoperations since several years ago. Their development can be classified into three categories of approach.

The first category focuses on haptics transmission system. In the field of control theory, (Ferrell, 1965) showed the transmission delay deteriorates the stability and the performance of remote manipulation and (Hannaford, 1989) opened up a new field in haptics. To stabilize a controller against unknown delay, a communication disturbance observer is proposed (Natori, 2008). In the field of applied technology, (Sato & Yakoh, 2000) implemented 1ms sampling loop as a network based control system, and (Oboe, 2001) realized web-based force feedback system. (Tsuji, 2004) evaluated the performance of a bilateral operation system between Slovenia and Japan approximately 9000km distance via the Internet. So the theory and the implementation technology have been researched well. However, it is still difficult to transmit keen haptic sensation through long delay or high jitter communication line.

The second one is about network QoS (quality of service) to improve the performance of information transmission systems. To reduce the jitter, several methods have been studied. Network delay consists of transmission delay, switching delay, queuing delay, and retransmission delay, and their magnitude vary greatly according to the network conditions (Gutwin et al., 2004). Some researches also focused on solutions based on network communication, either at a transport-layer, a network-layer, or an application-layer in the form of framing update messages (Shirmohammadi & Nancy, 2004) (Dodeller & Georganas, 2004). Some researches focused on the delay of transmission of position information of virtual object on Collaborative Virtual Environment (CVE) (IEEE 1995). The method of Synchronous Collaboration Transport Protocol (SCTP) sets the communication priority of information needed for synchronization of local and remote site (Boukerche et al., 2006). Thus, there are many researches to solve the problems of the network delay and the communication delay of the feedback information in the systems. However, there are no crucial solutions for the problems.

The third category is about the expression method to increase the accuracy of its operator's perception of remote site situation. Decorators were proposed to improve the operability of teleoperation system (Gutwin et al., 2004). Most of them have been studied in the field of CVE. They present supplementary information for teleoperation system. For example, the decorators indicate magnitude of delay, jitter, round-trip time, network lag and end-to-end delay by changing the color of pointing cursor (Gutwin et al., 2004) (Boukerche et al., 2006). They are useful because the operator can recognize the change of delay at a glance. Another decorator indicates the past tracks, the present position and predicted future position of the virtual object. Thus, decorators are different from the communicated environmental audio-visual information that indicates the situation of the remote site in teleoperation.

3. Decorators for Teleoperations

As mentioned in Section 1, a haptcs transmission system requires short delay communication line while a video transmission system requires wide bandwidth communication line. It is then natural to design the communication lines separately according to the required performance in such a teleoperation system. In this case, haptic information reaches first, and visual information appears last. Consequently, its operator may experience a strange feeling due to the time lag among those sensations. This is a problem targeted in this chapter.

This study proposes two decorators to overcome this issue. Though the word 'decorator' is derived from the field of CVE, the proposed decorators are different from the original ones. The proposed decorators are kinds of artificial cross-modal modification. One decorator, named visual decorator, generates a visual image of remote objects with haptic information and superimposes it onto the screen of a visual transmission system. Since haptic information reaches earlier than transmitted visual information, the decorator is fresher than the image of remote site. Additionally, the decorator image is synchronized to haptics transmission system in principle. So it is expected to resolve the strangeness of unsynchronized teleoperation system. Another decorator, named auditory decorator, generates artificial sound based on haptic information and mixes it with the sound of audio transmission system. Since auditory transmission system delays audio information artificially so as to realize lip sync, the audio decorator is also fresher than the sound of remote site. So it is expected to relax the strangeness too.

4. Teleoperation System

In this study, one task is set as the representative teleoperation. On the haptics transmission system (the detail is described in section 4.1), a stick is set as master and the rail is set as slave. They can rotate around a rotating shaft and the slave rail rotates following the master stick. So the operator controls the position of the rail by handling the master stick. Based on the system, a target task is controlling the position of a ball to stay close to the center of the rail. The ball is able to move freely on the rail. At the both ends of the rail, stoppers are set to prevent the ball from falling off the rail. Fig. 1 illustrates our setup

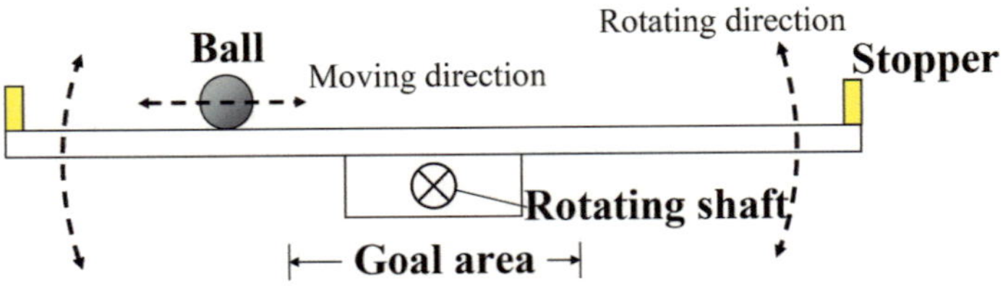

Fig. 1. Ball, slave rail and rotating shaft which are used for the target task

This task of controlling the position of the ball is designed as simples as possible so as to enable its operator achieve only with of audio, video or haptic sensations. The operator manipulates the stick by watching the ball and the rail, hearing the sound of the ball moving and perceiving the weigh of the ball. Moreover, even if the one or two kinds of information are lost, it is still possible to achieve the task with the other information. Thus, the task depends on the auditory, visual and haptic information, and these contributions are independent. That is why the task is suitable to set as the representative teleoperation. In this study, the rail is 950mm long. The ball is 32mm in diameter and the range of the goal area is within 40mm from the center of the rail.
For this task, an experimental tele-oepration system is constructed. Fig. 2 shows the overview of the teleoperation system. According to the above-mentioned background, auditory, visual and haptics transmission lines are set separately.

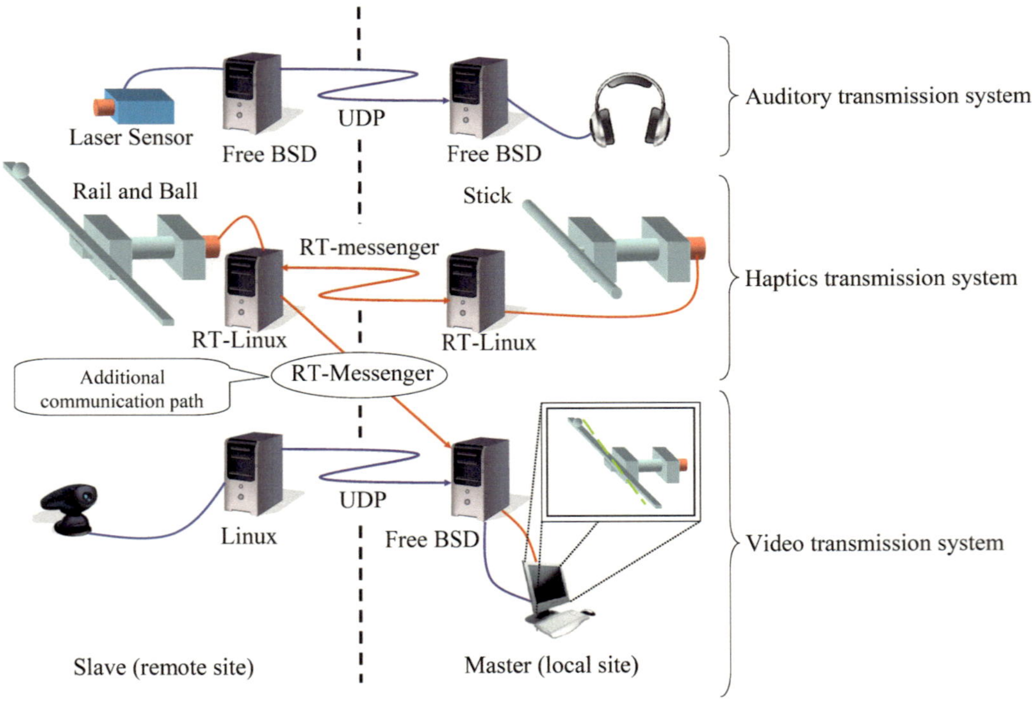

Fig. 2. Overview of experimental teleoperation system

4.1 Haptics Transmission System

Haptic information is transmitted through the center line in the Fig. 2 with real-time communication framework. The master stick at a local site and a slave rail at the remote site are controlled bilaterally (Iida & Ohnishi, 2004) so as to move as if those rotating shaft are connected directly. Because the haptics transmission system is controlled bilaterally, when the ball moves to the right of rail, the torque hanged on the master stick clockwise. Fig. 3 and 4 show the master stick and slave rail. Fig. 5 shows the implementation of the haptics transmission system.

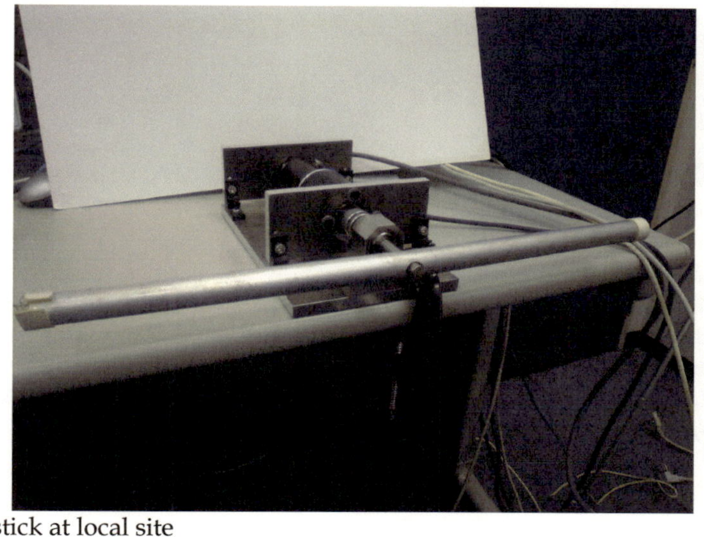

Fig. 3. Master stick at local site

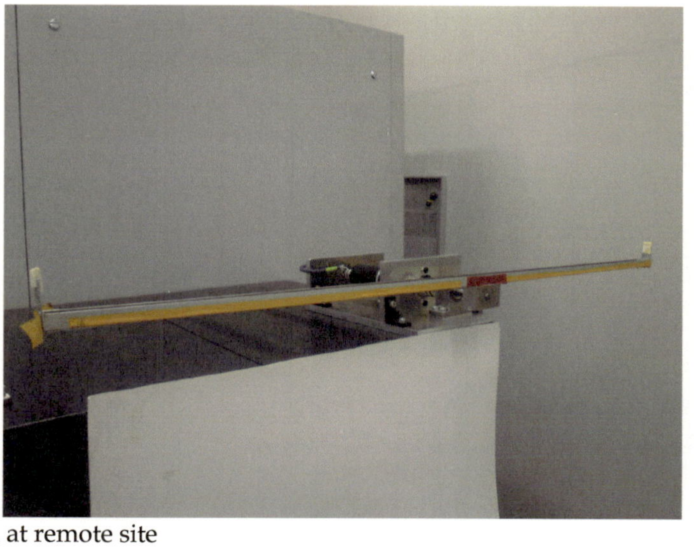

Fig. 4. Slave rail at remote site

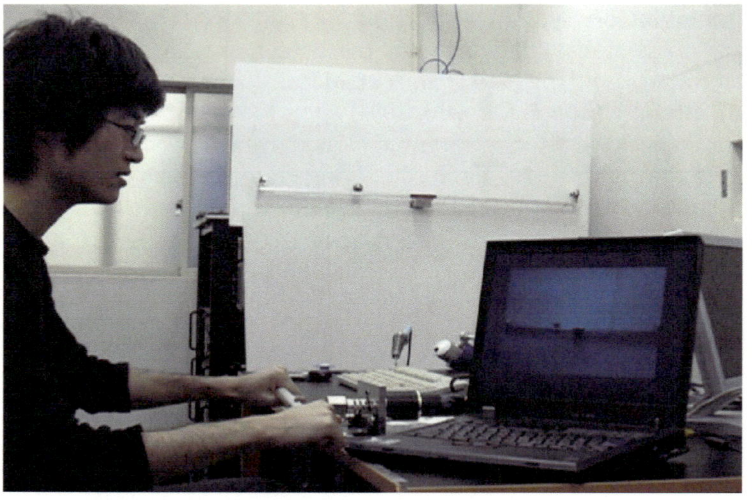

Fig. 5. Implementation of the haptics transmission system

RTLinux is used to support real-time operation of the haptics transmission system. In addition, RT-Messenger is used to enable the haptic system to communicate between local site and remote site with Ethernet protocol (Sato & Yakoh, 2000). The basic idea of RT-Messenger is to skip processing in software protocol stacks to minimize processing delay. This puts a packet directly into the head of hardware Tx queue in transmitting phase, and picks the packet directly from a Rx queue in receiving phase. Fig. 6 shows packet flow in transmission of RT-Messenger.

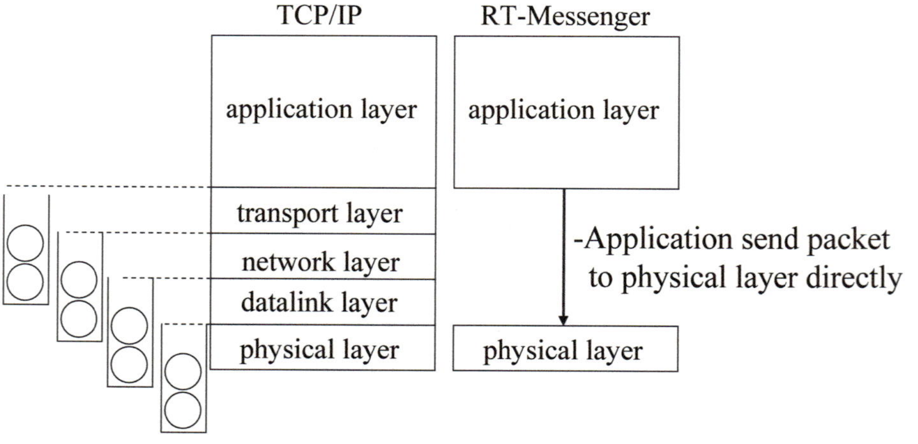

Fig. 6. Packet flow in transmission phase of RT-Messenger

Table 1 shows the specifications of the haptics transmission system.

Controller	
CPU	HT technology Pentium4 Processor 640 3.0GHz clock 2Mb cache
Motherboard	800MHz FSB
memory	1GB
NIC	Intel PRO/100 S Desktop Adaptor
OS	RTLinux-3.1 Linux Kernel Version 2.4.20

Control Device	
Motor, gear-head	MAXON MOTOR RE40,GP52C Torque constant 60.3 mNm/A Reduction ratio 12:1
Motor-driver	Servotechno PMA6 MAX. 110V
Rotary encoder	CANON R-1O 81,000 pulse per rotation
D/A board	Interface PCI-3345A Resolution 12bit Conversion time 10μs
Counter-board	Interface PCI-6204 Counter 32bit Maximum input frequency 1MHz

Table 1. Specifications of the haptics transmission system.

4.2 Video Transmission System (Live Video Streaming System)

For the feedback of the visual information, the live video streaming system is used. The lower communication line in Fig. 1 shows the system. This system displays 30fps video of the remote site to the master site with UDP protocol. Basically, the system is designed to minimize the processing delay (Endo et al., 2008). On top of this system, an artificial delay buffer is introduced to emulate orbitral communication delay of visual feedback. This buffer can emulate 120, 240, 360, 480ms delay. In the teleoperation system, the remote camera takes the image of remote slave and the local display shows the image. Table 2 shows the specification of the video transmission system.

Remote site	
CPU	HT technology Pentium4 Processor 640 3.0GHz clock 2Mb cache
Motherboard	800MHz FSB
memory	1GB
NIC	Broadcom NetXtreme 57xx Gigabit Ethernet Controller
VGA	NVIDIA Quatdro FX540
Camera	CIS CORPORATION VCC-8350CL Maximum resolution 60fps

Local site (monitor)	
Responsivity	8ms
Optimum resolution	1280*1024, Refresh rate 60Hz
Maximum resolution	1280*1024, Refresh rate 75Hz

Table 2. Specifications of the visual transmission system.

4.3 Auditory Transmission System

The auditory transmission system uses the upper communication line in Fig. 2. In the past research, ultrasonic sensors, which were located at the both ends of the rail measured the position of the ball and the system generated the sound according to the measurements as pseudo environmental sound. When the ball was far from the center of the rail, high frequency sine wave was given. Low frequency wave indicated that the ball was near the center. However, this method was not so effective in view of improving the operability of the teleoperation in the past experiments. In this time, therefore, the system does not provide the sound of the remote situation, but provides only additional sounds which indicate the position of the ball on the remote rail and helps the task.

5. Proposed Decorators

This study proposes visual decorators and auditory decorators which improve the operability of the teleoperation system.

5.1 Visual Decorators

This article assumes that delay of live video playout is longer than that of haptics transmission. With this assumption, the haptic information of remote site reaches to local site earlier than arrival of video information, i.e. haptic information is fresher than video information. So it is much worth visualizing haptic information. In this study, position and force information are transmitted as the haptic feedback. Based on these information, this study proposes two kinds of visual decorators which superimpose real-time remote site information on delayed remote video play backs. A position decorator shows the real-time CG image of the remote slave. In this study, the position decorator is bar and its rotation follows the remote rail. If the delay of haptic information is the same as the visual information, the graphic image will be overlapped perfectly with the remote image in the video play backs. However, since haptic information arrives earlier than video information, the position decorator indicates future position of the slave rail against the video playout. This decorator, thus, is expected to improve the operability of the teleoperation system especially when the delay of video play backs is long. This study also proposes a force decorator which visualizes the torque of the remote rail. Force itself is invisible physical abstraction. Therefore, there are many possibilities to visualize force information. This study simply uses bar-graph style appearance to visualize force information. In this study, when the ball is right side of the rail and the weight hangs on, the right bar-graph becomes long, and when it is on the left side, the left bar-graph grows.

The position decorator and force decorator are based on the haptic information. The information is already available to use because haptics transmission system transmit them simultaneously. As a result, haptic feedback system is modified so as to send position and force data also display site of live streaming system through RT-Messenger. In Fig. 2, the additional communication path is shown. SDL and OpenGL libraries were used to draw these decorators.

The effect of force decorator seems questionable compare to that of position decorator. So this article evaluates the effect of two visual decorators independently. Fig. 7 shows the implementation of the visual decorators.

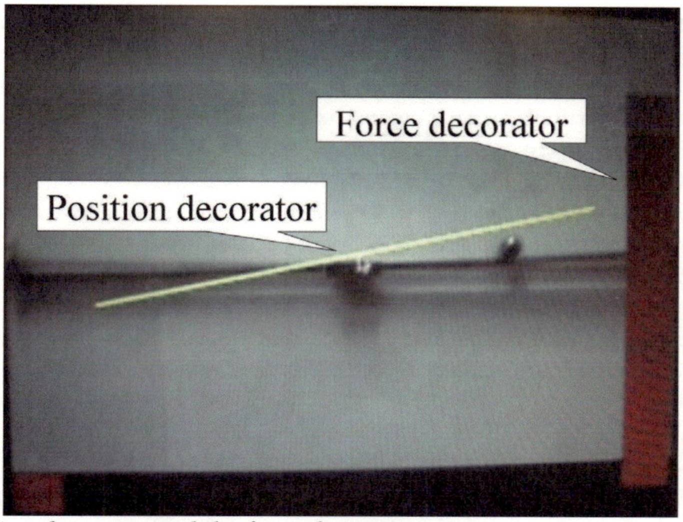

Fig. 7. The Position decorator and the force decorator

5.2 Proposed Auditory Transmission System and Auditory Decorator

The ball position is measured by a laser range finder. Fig. 8 shows tarrangement of mirrors and laser range finder. The box with a polymer reflector covered the ball, and two optical mirrors are located at the axis of the rotating shaft of the rail and the left end of it.

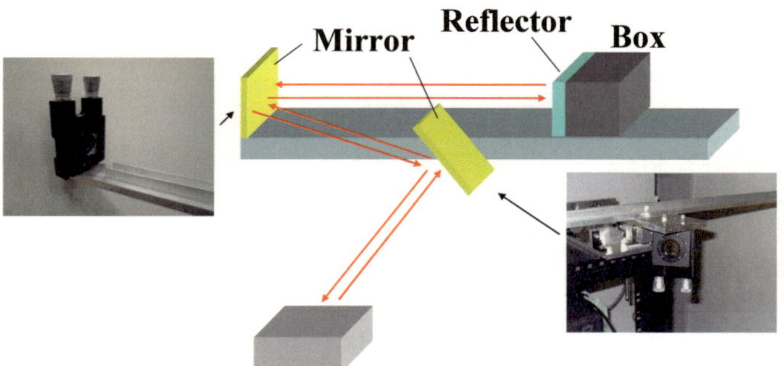

Fig. 8. Measurement of the weight position by laser sensor and optical mirrors

Table. 3 shows the specifications of the laser range finder.

Laser sensor	SICK DME5000
Sampling rate	2ms
resolution	1mm

Table 3. Specification of the laser range finder.

According to the measurements, the additional guiding sounds are generated. The sounds are nonverbal beep which can be perceived at once. In our past research, the pseudo sound indicates the ball position by the frequency this time. However, the method was not so effective to improve the operability of teleoperation. So the system varies the volume according to the ball position. The frequency of the sound is set 1000Hz. This is as much as that of siren, and human being can perceive easily the variation of the sound. Moreover, there is a context effect in our perceptual characteristics. The context effect is alternation in perception of the stimulus by the previous or next one. If the volume varies continuously, the volume is fluctuated by little and little when the ball moves slowly. In that case, the operator gets used to the sound by the context effect and hard to perceive the variation. So the volume variation of the guiding sound should be varied discretely for the operator to perceive the variation and the position of the ball easily. This system sets 4 levels of volume. When the ball is close to the end of the rail, the guiding sound is generated at the loudest volume. The minimum volume indicates that the ball is near the center of the rail, and when the ball is in the goal area, the sound is off. In addition, human being has the ability of sound localization which is used recognize the direction of the sound source. Based on the ability, in case the ball is right side on the remote rail and the stick is rotated clockwise, the right speaker gives the tone and other may around. Fig. 9 shows the correspondence of the indicating position and the volume of additional guiding sound.

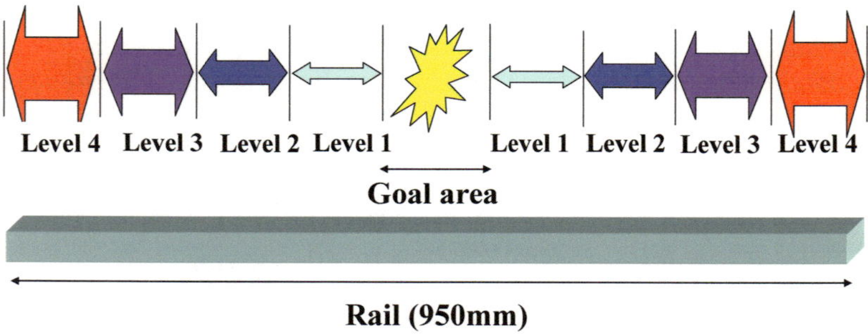

Fig. 9. Correspondence of the indicating position and the volume of additional guiding sound in the auditory feedback system

In addition to the auditory feedback system, auditory decorator is implemented. When the ball is in the goal area, the auditory decorator gives the other beep sound. Its frequency is higher than that of guiding sound of auditory feedback system.

6. Experiments and Results

To evaluate the operability of teleoperation system and the efficiency of the proposed decorators, task achievement time is measured under several conditions.

As mentioned in section 4, controlling the position of the ball to be center of the rail is set as the target task. The initial time, the ball is put on the left end of the rail and the operator starts to manipulate the stick. When the ball locates in the goal area of the rail stably, the operation is finished. The elapsed time is used as an index of usability, i.e. shorter operation time, higher usable condition. Artificial delay time of the live video streaming and the combination of used decorators are used as condition parameters. Delay time are selected as 120, 240, 360, and 480ms.

6.1 Evaluation of Visual Decorators

To examine the stability of the operation, the reaction force of the remote rail is measured in the experiments of visual decorators. When the ball is far from the center of the rail, the torque becomes large because of the weight of it, and the reaction force becomes large. If the reaction force is small, the ball is near the center and the operation is stable. Used combinations of decorators are position and force (P+F), position only (P), force only (F), and no decorators (Non). For each sixteen combinations of all video delay and decorators, 5 participants operated the task twice, and averages of 10 times of measurements are used as the usability index of each condition.

A. results of the task achievement time

Table 4 shows the task achievement times of the experiments of visual decorators

	Combination of decorators			
Video delay [ms]	Non [s]	P[s]	F[S]	P+F [s]
120	33.90	21.43	27.46	32.5
240	38.75	30.62	35.87	32.1
360	50.00	37.46	41.81	42.5
480	56.62	36.53	50.84	44.8

Table 4. Average task achievement time for each experimental circumstance of visual decorators

From table 4, no decorator use, the most left column, shows the longest value for each row. From this result, the decorators were effective at any values of the delay. Especially, the position decorator was very effective and shortened the task achievement time. The position decorator indicated how the operation of the master stick is followed by the manipulation of the remote rail, and operator may understand that at glance. The result in 120ms shows that position decorator was effective. However, the effect is less than it in case of the delay is long. When the delay is 480ms, the task achievement time can be shortened for 20 seconds compared to the case without decorators. To locate the ball stably, the operator manipulated little by little while taking the feedback of the action of the rail. Therefore, the difference between the position information of the rail in the video playout and the bar of position decorator was small when the delay of the playout was short. Therefore, the position decorator is less effective when the delay is short.

When the delay of the visual feedback is set as 120ms with position decorator, the task achievement time shows the shortest value. On the other hand, when the delay is 480ms without decorators, the time indicates the longest value.

In cases of the delay is 480ms, regardless of presence and the combination of the decorators, the task achievement time has grown. The examinee said the operability is worsened extremely in the cases. Also, it is said that the examinee got used to operate under the other three delay values, 120, 240 and 360ms.

Next, the experimental result of force decorator is discussed. The force decorator cannot shorten the task achievement time significantly compare to the position decorator achieves. It means that the force decorator is less effective than position decorator in the teleoperation. The purpose of force decorator is to show the position or movement of the ball on the remote rail from the torque information, and to indicate to which site of the torque is large. However, the examinee says that it is difficult to pay attention to the moving of force decorator besides the slave on the monitor because they concentrate on the action of the slave rail. In addition, it is not useful to visualize the force infomation to shorten the task achievement.

B. stability of the operation

To mention the stability of the operation, this article considers the results of the reaction force of the remote site by the view point of the combinations of the decorators.

Fig. 10, 11, 12, 13 show the representative results of the reaction force and the achievement time of each combinations of the decorators.

When no decorators are used or the force decorator is used, the reaction force fluctuated greatly in the cases of the video delay are 360ms and 480ms. It is because that the manipulation of master became large by not stabilizing the position of the ball and the operation.

These results show that the force decorator is not able to give the positive effect of improving the operation about 20 seconds from the beginning of the task in cases of the delay are 360 and 480ms. However, the reaction force converged after the state which the reaction force and the position of the ball are not steady (for example, II in the Fig. 11 shows the state). This is because the movement of the ball is not steady, the bar-graph of force decorator became long and the effect was appeared. So, the reaction force converged after 20 seconds from the beginning of the task. Force decorator is effective when the operation is not stable.

On the other hand, when only the position decorator is used, the swinging of the reaction force is small (for example, I in Fig. 10 shows the swinging range). As a whole regardless of the delay of live video streaming, the operations can be done stably. Especially, steady operations are done when the delay are 240, 360, 480ms compared with the cases that decorator are not used, and the effectiveness of position decorator is shown.

When two decorators are used, the operations are stable. These results suppose that the position decorator helped the teleoperation

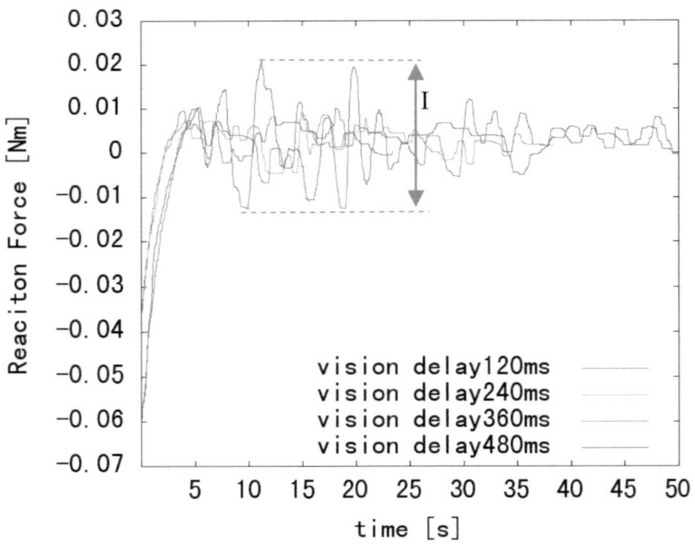

Fig. 10. Reaction force response when no decorators are provided

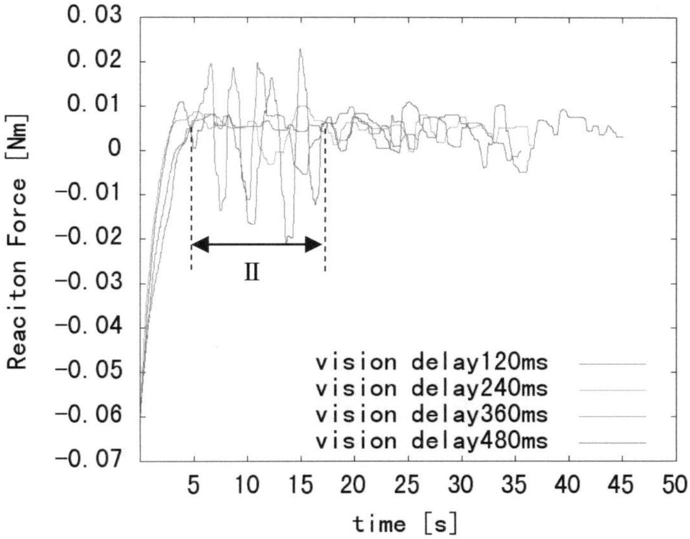

Fig. 11. Reaction force response when force decorator is provided

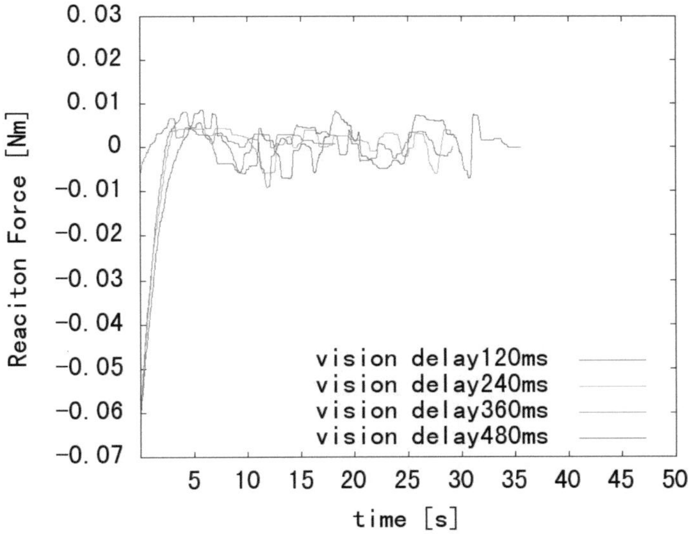

Fig. 12. Reaction force response when position decorator is provided

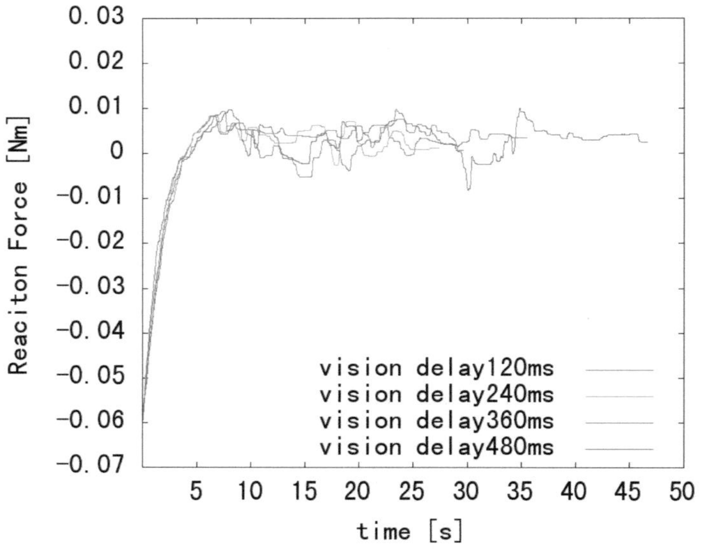

Fig. 13. Reaction force response when position decorator and force decorator are provided

6.2 Evaluation of Auditory Decorators

In the experiments which evaluate the proposed auditory transmission system and decorator, three kinds of experimental conditions are set. No auditory feedback (Non), with auditory transmission system and no decorator (T), with auditory transmission system and decorator (T+D). For each twelve combinations of all video feedback delay (120ms, 240ms, 360ms and 480ms) and auditory feedback, 5 participants operated the target task.
Table 5 shows the task achievement times of the experiments of auditory transmission system and decorator.

Video delay [ms]	Non [s]	F [s]	F+D [s]
120	31.4	33.0	32.5
240	41.3	35.2	32.1
360	45.1	43.9	42.5
480	54.7	44.8	44.8

Table 5. Average task achievement time for each experimental combination of auditory transmission system and decorator

Table 5 shows that the averages are rarely different in case the video playout delay is 120ms. It is believed that the video delay did not cause the deterioration of the operability of the teleoperation in that case. Other results show that proposed auditory transmission system and decorator shorten the task achievement time by 9-24%, and it means that the results are independent from whether the auditory decorator is provided or not.
Moreover, the task can be done without video playout, but with the haptics and auditory transmission systems. The average of the task achievement time without video playout is 69.7 s.

7. Conclusion

This study proposed to introduce decorators to overcome the difficulty of teleoperation system through delayed communication line. The proposed visual decorators indicated some information superimposed on delayed video playout. Since the delay of information decorator used is shorter than that of video, decorator can indicate future information over video playout. This future information is expected to improve the operability of teleoperation system. The experimental results showed that the proposed decorators are useful in the case that delay is long. Especially, position decorator showed significant effect to the teleoperation system and improves the operability. Position decorator rotates like the stick and rail, so it is easy to recognize the information given by position decorator for operator. On the other hand, the action of force decorator was different from that of the experimental task. In addition, the task requires much concentration. Therefore, the operator cannot pay attention to force decorator. Decorator indicates the supplemental information for the teleoperation system and encourages the operator to manipulate properly. Thus, it needs its operator to manipulate immediately for the problems of the network delay and the communication delay of the feedback information. To do the immediate manipulation, the visual information is the most effective. In the view point of "immediacy", it is important whether the optical cue can be recognized easily and user can reflect the information from it to the manipulation momentarily. Therefore, the visual information like position is effective to use as decorator because it is easy to recognize and understand. That is why, position decorator is much useful than force decorator. The visual decorators which are optical cue are useful to the real object and the teleoperation system, and especially, the decorator which indicates the position information of the object is useful.

The proposed auditory transmission system and decorator also shorten the task achievement time, and improve the operability of the teleoperation.

The conclusion of this study is that the manipulation aid by visualizing the haptic information and the guiding additional tone are effective, thus the decorators helps teleoperation. As the future work, designing the other kinds of decorators and discussion which methods is much effective should be done. In addition, the quantitative evaluation of the effects of the decorators should be proposed.

Acknowledgments

This work was supported by Grant-in-Aid for Scientific Research (B)(20300079): matching fund subsidy from JSPS (Japan Society for the Promotion of Science). Hirotaka Sugiyama and Kenji Takahashi assisted set up the auditory decorator.

8. References

Boukerche, A.; Shirmohammadi, S. & Hossain, A. (2006). Moderating Simulation Lag in Haptic Virtual Environments. *Proceedings of IEEE the 39th Annual Simulation Symposium,* pp. 269-277.

Dodeller, S. & Georganas, N. D. (2004). Transport Layer Protocols for Tele-haptics Update Message. *Proceedings of Biennial Symposium on Communications.*

Endo, K.; Yoshida, K. & Yakoh, T. (2008). Low Delay Live Video Streaming System for Interactive Use. *Proceedings of IEEE International Conference on Industrial Informatics,* pp.1481-1486.

Ferrell, W. D. (1965). Remote Manipulation with Transmission Delay. *IEEE Trans. Human Factor in Electronics,* Vol.6, pp.24-32.

Gutwin, C.; Benford, S.; Dyck, J.; Fraser, M. ; Vaghi, I. & Greenhalgh, C. (2004). Revealing Delay in Collaborative Environments. *Proceedings of ACM Computer Human Interaction,* pp.503-510.

Hannaford, B. (1989). A Design Framework for Teleoperators with Kinesthetic Feedback. *IEEE Trans. Robotics and Automation,* Vol.5, Issue 4, pp.426-434.

IEEE. (1995). IEEE Standard for Distributed Interactive Simulation and Application Protocols. *Proceedings of IEEE Standard 1278-1995*

Iida, W. & Ohnishi, K. (2004). Reproducibility and Operationality in Bilateral Teleoperation. *Proceedings of IEEE International Workshop on Advanced Motion Control,* pp.217-222

Nardi, B.; Schwarz, H.; Kuchinsky, A.; Leichner, R. ; Whittaker, S. & Scalabassi, R. (1993). Turning Away from Talking Heads: The Use of Video-as-Data in Neruosurgey. *Proceedings of INTERCHI,* pp.327-334

Natori, K. & Ohnishi, K. (2008). A Design Method of Communication Disturbance Observer for Time-Delay Compensation, Taking the Dynamic Property of Network Disturbance Into Account. *IEEE Trans. Industrial Electronics,* Vol.55, Issue 5, pp.2152–2168.

Oboe, R. (2001). Web-Interfaced, Force-Reflecting Teleoperation Systems. *IEEE Trans. Industrial Electronics,* Vol.48, Issue 6, pp.1257-1265.

Sato, H. & Yakoh, T. (2000). A Real-Time Communication Mechanism for RTLinux. *Proceedings of IECON 2000 Annual Conference on the IEEE Industrial Electronics Society,* Vol.4, pp.22-28.

Shirmohammadi, H. & Nacy, H. W. (2004). Shared Object Manipulation with Decorators in Virtual Environments. *Proceedings of IEEE International Symposium on Distributed Simulation Real-Time Applications,* pp.230-233.

Tsuji, T.; Kato, A.; Ohnishi, K.; Hace, A. & Jezernik, K.; (2004). Safety Control of Teleoperation System under Time Varying Communication Delay. *Proceedings of AMC 2004 IEEE International Workshop on Advanced Motion Control,* pp.463-468.

3

Predictor Based Control Strategy for Wheeled Mobile Robots Subject to Transport Delay

Alejandro Alvarez-Aguirre
Eindhoven University of Technology
Eindhoven, The Netherlands

1. Introduction

The performance and stability of a control system can be directly affected by a time-delay located either in its input, output, or both. In the case of a mobile robot, an input time-delay may become critical in different situations, such as when vision is used as the localization technique and a high frame per second rate is demanded, or when centralized control of multiple agents is desired, or even if very accurate regulation or tracking performance is required. It turns out that the control laws derived from the mathematical models that include an input time-delay are of a noncausal nature, thus requiring some sort of state prediction or estimation in order to implement them.

Initially, this work considers the control of a wheeled mobile robot (unicycle-type or omnidirectional) which is subject to an input time-delay. The causality problem involved in the proposed solution is tackled by considering the nonlinear case of the well known Smith predictor compensator. By doing this, the implementation of a noncausal feedback is possible and the system's performance improves under an input-time delay. It is worth noting that the necessity to consider this type of time-delay has motivated the use of discrete-time models that allow the analysis of the time-delay's effects and the synthesis of discrete-time controllers designed to compensate such effects. The complexity of the problem increases especially when the time-delay is included in the model due to the nonlinear nature of mobile robots.

Additionally, this work also proposes an extension that is able to cope with bilateral time-delays and which is based on the solution of the input time-delay problem affecting a mobile robot mentioned previously. Although rather simple, this extension is of particular importance due to the fact that it brings the proposed control strategy closer to the realm of telecontrol and teleoperation. The possibility to consider a mobile robot affected by a network time-delay rather than only an input time-delay opens a whole new range of possibilities and applications in which remotely controlling this kind of devices is possible.

In Section 2 of this chapter the classical Smith predictor is introduced together with its extensions for continuous and discrete time nonlinear systems. In the following section, the continuous time posture kinematic model of a unicycle-type and an omnidirectional mobile robot are presented. These models consider an input-time delay and are used to derive their exact discrete time posture kinematic models, which give way to the proposition of appropriate tracking control laws. Due to the time-delay, both control laws are found to be noncausal and thus, the requirement of a prediction scheme is concluded. The integration of the discrete time prediction scheme and the noncausal control laws is carried out in Section 4 and an extension that copes with bilateral time-delays is proposed. Numerical simulations are presented in Section 5, while conclusions and recommendations for future work are provided in Section 6.

2. The Smith Predictor

The Smith predictor compensator was first proposed by (Smith, 1957) and constitutes one of the simplest methods to control a Single-Input Single-Output (SISO) stable linear system with an input time-delay. Throughout the years, several modifications and extensions have been proposed to this compensator in order to accommodate, among others, Multiple-Input Multiple-Output (MIMO) linear systems, nonlinear systems, and robustness and disturbance rejection requirements. The main idea behind these *Smith predictors* is the use of a controller structure which extracts the time-delay out of the control loop and allows a feedback design based on a delay free system, (Michiels and Niculescu, 2007). In other words, the Smith predictor compensator enables the prediction of the states of a system at a given time instant in the future. The magnitude of the prediction time for the states is limited by the magnitude of the time-delay affecting the system. This prediction allows the implementation of a non-causal control law which can be used to control a system when it is subject to an input time-delay.

2.1 Linear Smith Predictor

This subsection is based on the work of (Kravaris and Wright, 1989), in which the working principle behind the Smith predictor is analyzed and where its configuration for the state space representation is explained.

Consider a linear SISO system in its state space representation and subject to an input time-delay τ,

$$\dot{x}(t) = Ax(t) + Bu(t-\tau), \tag{1a}$$

$$\dot{y}(t) = Cx(t), \tag{1b}$$

where $x(t)$ constitutes the system's state vector of dimension n, $u(t-\tau)$ its time-delayed input, $y(t)$ its output, A is an $n \times n$ constant matrix, B is a constant vector of dimension $n \times 1$ and C is a constant vector of dimension $1 \times n$.

The implementation of the Smith predictor for system (1) is subject to the following propositions:

Proposition 2.1 *System (1) is open-loop stable.*

Proposition 2.2 *The delay free part of system (1), as expressed in (2), has stable zero dynamics;*

$$\dot{\xi}(t) = A\xi(t) + Bu(t), \tag{2a}$$

$$y(t) = C\xi(t), \tag{2b}$$

where $\xi(t)$ is the state vector of the delay free part of the system.

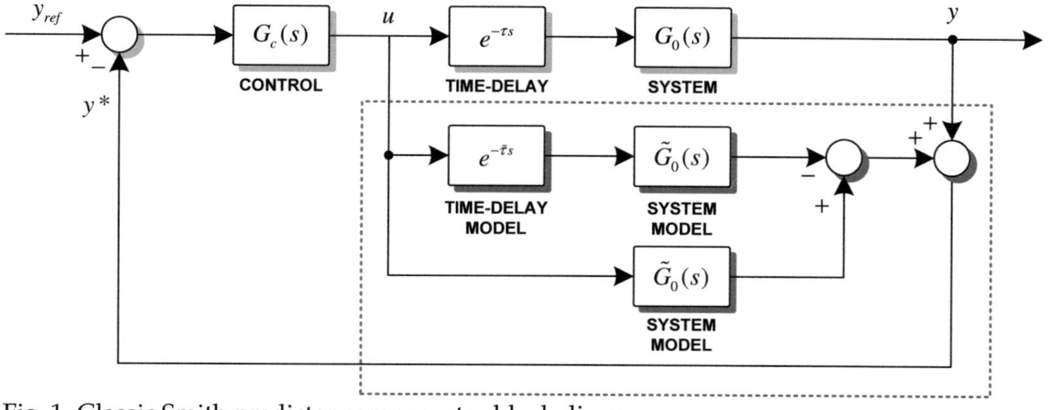

Fig. 1. Classic Smith predictor compensator block diagram.

Remark 2.3 *Note that in almost every variation and extension of the Smith predictor Proposition 2.1 turns out to be fundamental and therefore can not be discarded when considering nonlinear systems. Moreover, Proposition 2.2 is also crucial for nonlinear controllers, including the input/output linearization technique, as explained in (Kravaris, 1987) and (Kravaris, 1988).*

The transfer function of system (1) is given by,

$$\frac{y(s)}{u(s)} = G_0(s)e^{-\tau s}, \tag{3}$$

where all the poles and zeros of $G_0(s)$ are located on the left hand-plane. The classical block diagram structure of the Smith predictor for this system is shown in Fig. 1.

In the figure, $\tilde{G}_0(s)$ and $\tilde{\tau}$ represent the system's and time-delay's model respectively, and y_{ref} constitutes the reference signal. The main idea behind the Smith predictor is to compute the difference between the delay free and delayed model of the system. The signal that results from the predictor is added to the system's measured states in order to predict their value considering no time-delay is present. This predicted output y^* then enters the controller $G_c(s)$. The closed-loop transfer function resulting from the block diagram in Fig. 1 yields,

$$\frac{y(s)}{y_{ref}(s)} = \frac{G_c(s)G_o(s)}{1 + G_c(s)\tilde{G}_o(s) + G_c(s)\left(G_o(s)e^{-\tau s} - \tilde{G}_o(s)e^{-\tilde{\tau}s}\right)}e^{-\tau s}. \tag{4}$$

When the models of the system and the time-delay considered in the predictor are perfect, i.e. when $\tilde{G}_o(s) = G_o(s)$ and $\tilde{\tau} = \tau$, the closed-loop transfer function (4) becomes,

$$\frac{y(s)}{y_{ref}(s)} = \frac{G_c(s)G_o(s)}{1 + G_c(s)G_o(s)}e^{-\tau s}. \tag{5}$$

The structure of the closed-loop transfer function (5) and the interpretation already given to the feedback signal y^* indicates that the parametrization of the controller $G_c(s)$ should be in terms of the delay free part of the model, i.e. $G_o(s)$. Moreover, it is worth noting that the characteristic equation of system (5) is,

$$1 + G_c(s)G_o(s) = 0. \tag{6}$$

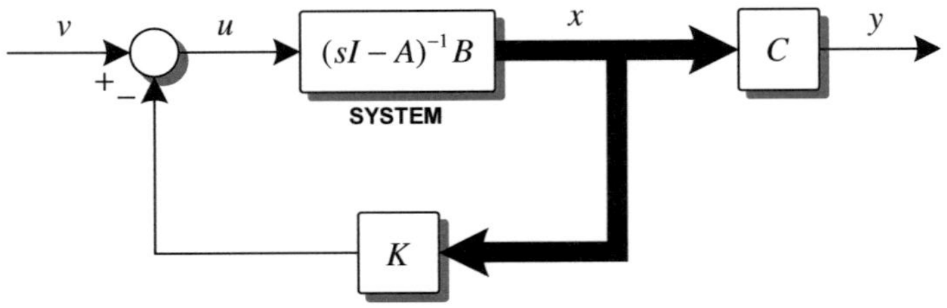

Fig. 2. Linear system in state space representation.

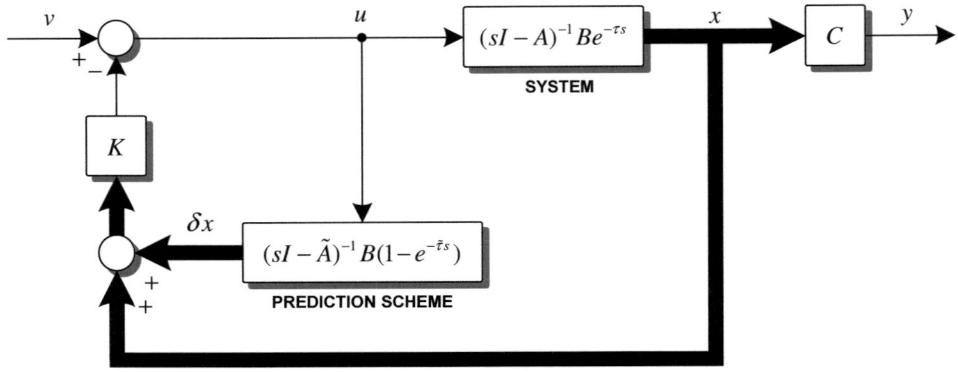

Fig. 3. Linear Smith predictor compensator in state space representation.

The concepts presented so far can be directly translated to the state space representation of a linear system (1). This, in turn, will simplify the extension of the idea behind the Smith predictor to nonlinear systems.

Consider now that the system in the state space representation is not affected by a time-delay, i.e. $\tau = 0$. Moreover, the system is subject to a static feedback of the type $u(t) = v(t) - Kx(t)$, where $v(t)$ is the system's reference and K is a group of gains. A block representation depicting this structure is shown in Fig. 2.

The system's closed-loop transfer function is given by,

$$\frac{y(s)}{v(s)} = \frac{CAdj(sI-A)B}{det(sI-A) + KAdj(sI-A)B}. \tag{7}$$

If the system is now subject to a time-delay, i.e. $\tau \neq 0$, a similar structure to the classical Smith predictor can be proposed for the state space representation. In this case, the value of the system's states if no time-delay was present can be obtained by adding a corrective signal to the states measured from the output. Such signal can be constructed by computing the difference between the delayed and the delay-free states. The prediction scheme for a system in the state space representation is depicted in Fig. 3.

Considering once more a perfect modeling of the system and the time-delay, i.e. $\tilde{A} = A$, $\tilde{B} = B$ and $\tilde{\tau} = \tau$, the closed-loop transfer function of the block diagram in Fig. 3 becomes,

$$\frac{y(s)}{v(s)} = \frac{CAdj(sI - A)B}{det(sI - A) + KAdj(sI - A)B} e^{-\tau s}. \tag{8}$$

It is worth noting that expressions (7) and (8) are the same except for the time-delay term $e^{-\tau s}$. It should be clear that nothing can be done with this term and that a non-causal feedback is required in order to obtain a system's response which is not time-delayed.

From the closed-loop transfer function (8) it should also be clear that this structure allows to select the closed-loop poles of the time-delayed system by using some kind of pole placement technique on the delay free part of the system. It is worth noting that (Morari, 1989) explains that if stability is the only criterion considered, the previous statement is correct. However, if the system's performance is the criterion taken into account, several considerations should be made in order to correctly tune the controller, see (Morari, 1989).

In summary, the problem of pole placement for a system subject to a time-delay can be reduced to the problem of pole placement for the delay free part of the system if a Smith-type predictor is embedded into the system's control structure.

The basic properties of the Smith predictor have been extensively explained in (Jerome and Ray, 1986) and apply to the different prediction schemes derived from it. It is worth noting that these properties assume a perfect system and time-delay model. In order to provide further insight into the behavior of the Smith predictor, these properties can be very briefly stated as:

Property 1. The Smith predictor eliminates the time-delay of the closed-loop characteristic equation of a system.

Property 2. For changes in the system's operating point, the Smith predictor provides the controller with an immediate prediction τ units of time into the future of the effects of its control action on the system's predicted output.

Property 3. The structure of the Smith predictor implicitly divides the model of the dynamical system in two parts or terms. The first one being the time-delay $e^{-\tau s}$ and the second one the remaining system dynamics $G_0(s)$. This two terms can then be treated in a completely independent way.

2.2 Nonlinear Smith-Type Predictor

This subsection is based on the work of (Kravaris and Wright, 1989), in which the Smith predictor in its state space representation is extended in order to accommodate continuous time nonlinear systems.

Consider the classic representation of a delay free, continuous time nonlinear system,

$$\dot{x}(t) = f(x(t)) + g(x(t))u(t), \tag{9a}$$

$$y(t) = h(x(t)). \tag{9b}$$

where $u \in \mathbb{R}^m$ is the system's input, $x \in \mathbb{R}^n$ is the system's state, and $y \in \mathbb{R}^m$ is the system's output.

It is assumed system (9) satisfies Propositions 2.1 and 2.2 (for a delay free system). Moreover, if the system has a relative degree vector $r_1,...,r_m$ at point x^0, then following static state feedback,

$$u(t) = \Psi(x(t),v(t)) = \frac{1}{L_g L_f^{r-1} h(x(t))} \left(v(t) - L_f^r h(x(t)) \right), \tag{10}$$

with the auxiliary control $v(t)$ defined as,

$$v(t) = y_R^{(r)} - \sum_{i=1}^{r} c_{i-1} \left(L_f^{(i-1)} h(x(t)) - y_R^{(i-1)} \right), \tag{11}$$

will cause the system to follow the reference $y_R^{(r)}$, i.e. the input/output behavior of the system satisfies the expression,

$$y_R^{(r)} = v(t). \tag{12}$$

In order to guarantee input/output stability, the control parameters c_{i-1} should be selected in such a way that all the poles of the resulting linear subsystem are located in the left hand-plane.

Consider now a nonlinear system with an input time-delay and which satisfies Propositions 2.1 and 2.2,

$$\dot{x}(t) = f(x(t)) + g(x(t))u(t-\tau), \tag{13a}$$
$$y(t) = h(x(t)). \tag{13b}$$

Concerning the extension of the concept of input/output linearization for system (13), it is worth noting that, due to the fact that the system is subject to an input time-delay, it is not possible to find a causal static feedback which transforms the system into a linear delay free one by using the methodology presented so far. The best that can be done is achieve a linear input/output behavior which is time-delayed, given by,

$$y_R^{(r)} = v(t-\tau). \tag{14}$$

A state feedback with a predictive action similar to the Smith predictor may now be considered. In order to achieve this, the state space representation of the Smith predictor presented in Subsection 2.1 will be used.

The system's model can be used to compute a corrective signal which, when added to the measured states, predicts their values if no time-delay was present. With this, the predicted states can be fed into the controller by means of a static feedback Ψ. A block diagram of the resulting structure is shown in Fig. 4.

The nonlinear predictor may be characterized by the following expressions,

$$\dot{\tilde{x}}(t) = \tilde{f}(\tilde{x}(t)) + \tilde{g}(\tilde{x}(t))u(t), \tag{15a}$$
$$\dot{\hat{x}}(t) = \tilde{f}(\hat{x}(t)) + \tilde{g}(\hat{x}(t))u(t-\tilde{\tau}), \tag{15b}$$
$$\delta x(t) = \tilde{x}(t) - \hat{x}(t), \tag{15c}$$

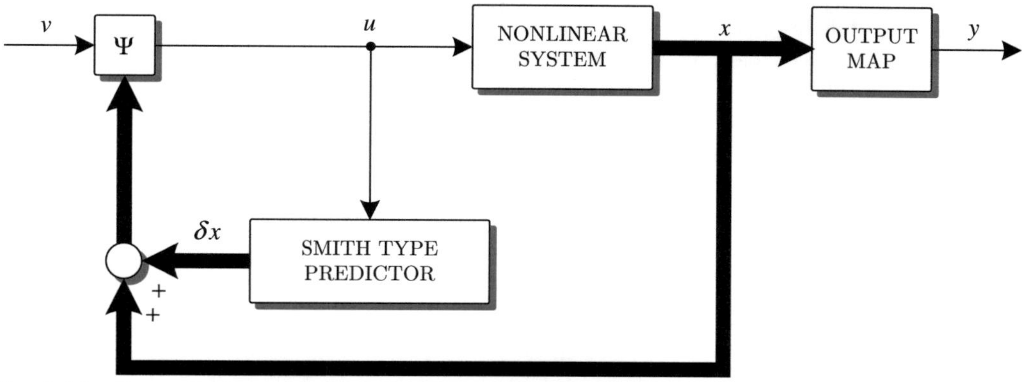

Fig. 4. Nonlinear Smith-type predictor compensator.

where $\tilde{x}(t)$ represents the states of the delay free model of the system, $\hat{x}(t)$ represents the states of the delayed model of the system, and $\delta x(t)$ represents the predictor's output, composed by computing the difference between the two models. This produces, in a similar way to the linear case, the following term entering the controller,

$$x^*(t) = x(t) + \delta x(t). \tag{16}$$

A perfect modeling of the system and the time-delay, i.e $\tilde{f} = f, \tilde{g} = g, \tilde{\tau} = \tau$ and $\hat{x}(t) = x(t)$, results in $x^*(t) = \tilde{x}(t)$. This means that the controller is actually fed by $x^*(t) = \tilde{x}(t)$, which in reality is the state of the delay free model of the system and in fact constitutes the prediction of the system's state τ units of time into the future, i.e. $x(t+\tau)$.

The predicted state together with the static state feedback Ψ, given by equation (10), generates the following input/output behavior of the delay free model of the system,

$$y_R^{(r)} = v(t). \tag{17}$$

Considering this, the feedback law will now be given by,

$$u(t) = \frac{1}{L_g L_f^{r-1} h(x(t+\tau))} \left(v(t+\tau) - L_f^r h(x(t+\tau)) \right). \tag{18}$$

As the control signal (18) experiences a time-delay, the system's input becomes,

$$u(t-\tau) = \frac{1}{L_g L_f^{r-1} h(x(t))} \left(v(t) - L_f^r h(x(t)) \right), \tag{19}$$

and therefore the system will track a delayed version of the reference signal.

2.3 Discrete-Time Nonlinear Smith-Type Predictor

The extension of the Smith predictor to discrete time nonlinear systems was carried out by (Henson and Seborg, 1994) and considers the discrete time model of a nonlinear system as,

$$x(k+1) = f(x(k)) + g(x(k))u(k-\tau), \tag{20a}$$

$$y(k) = h(x(k)). \tag{20b}$$

The only consideration that has to be made in order to obtain a predictor for discrete time nonlinear systems is that the modification of the Smith predictor carried out in Subsection 2.2 has to be derived in terms of the discrete time model (20).

For instance, consider that the future values of the state in model (20) are required for time instant $k+i$, i.e. i time instants into the future. Such value would be given by,

$$x(k+i) = f(x(k+i-1)) + g(x(k+i-1))u(k+i-\tau-1). \tag{21}$$

On the other hand, considering a prediction for time instant $i = \tau$, the output of the predictor presented in (15) can be expressed in discrete time based on expression (21), i.e.,

$$\breve{x}(k+1) = \tilde{f}(\breve{x}(k)) + \tilde{g}(\breve{x}(k)u(k)), \tag{22a}$$

$$\hat{x}(k+1) = \tilde{f}(\hat{x}(k)) + \tilde{g}(\hat{x}(k)u(k-\tilde{\tau})), \tag{22b}$$

$$\delta x(k) = \breve{x}(k) - \hat{x}(k). \tag{22c}$$

Once more, a perfect modeling of the system and the time-delay, i.e $\tilde{f} = f, \tilde{g} = g, \tilde{\tau} = \tau$ and $\hat{x}(k) = x(k)$, results in $x^*(k) = \breve{x}(k)$. In other words, the controller is actually being fed with $x(k+\tau)$ and a noncausal control law may be implemented.

The previous results are summarized in two properties by (Henson and Seborg, 1994),

Property 1. If a perfect model of the system and the time-delay are used, the controller will receive the signal $x^*(k) = x(k+\tau)$ for all $k \geq 0$.

Property 2. If the closed-loop system is asymptotically stable, then $x^*(k) = x(k+\tau)$ in the limit as $k \to \infty$.

The cited work also explains the reasons why the proposed predictor may yield poor state predictions when mismatch between the model and the system exists or when unknown perturbations affect the system. The latter applies for both, the continuous and discrete time case.

3. Wheeled Mobile Robots (WMR)

A mobile robot may be defined as an electromechanical device which is capable of displacing within its workspace and can be classified according to its type of locomotion, e.g. by means of legs, wheels or tracks. A fundamental issue when considering the analysis, design, implementation and control of wheeled mobile robots (WMR) is precisely their type, layout, configuration and characteristics. For example, the wheels of a mobile robot may be conventional or omnidirectional, and of fixed or adjustable orientation. Moreover, the number, type and layout of the wheels of a WMR determines its classification and number of degrees of freedom. A practical mobile robot moving on a plane should have as minimum two degrees of freedom and as maximum three (Canudas De Wit et.al., 1996). This work features a unicycle-type mobile robot or type (2,0), which possesses two degrees of mobility provided by a translational and a rotational velocity. Also included is an omnidirectional mobile robot or type (3,0), which possesses three degrees of mobility provided by a rotational velocity and two linear ones.

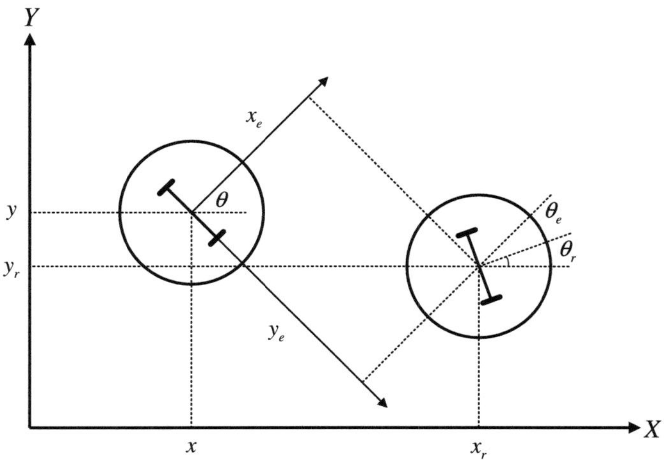

Fig. 5. Unicycle-type mobile robot and error coordinates.

3.1 Posture Kinematic Model

In general, the mathematical model of a WMR is nonlinear and, in cases such as the unicycle-type mobile robot, may even belong to the class of systems denoted as non-holonomic, which are characterized by non-integrable restrictions in their velocities. The posture kinematic model of a mobile robot provides as output specific information about the location and orientation of the vehicle within its workspace and uses the robot's velocities as inputs. In particular, the discrete-time posture kinematic model of a mobile robot allows a closer control of the sampling time at which information is sent and received from the vehicle.

3.1.1 Unicycle-Type Mobile Robot

The kinematic model of a unicycle-type mobile robot can be easily derived by considering the geometric representation given in Fig. 5. The velocity components with respect to the Cartesian coordinate system $X - Y$ are obtained as in (Canudas De Wit et.al., 1996), (Campion et.al, 1996), i.e.,

$$\dot{x}(t) = v(t)\cos\theta(t), \tag{23a}$$

$$\dot{y}(t) = v(t)\sin\theta(t), \tag{23b}$$

$$\dot{\theta}(t) = \omega(t), \tag{23c}$$

in which $x(t)$ and $y(t)$ denote the robot's position in the workspace w.r.t. the coordinate frame X-Y, $\theta(t)$ corresponds to its orientation with respect to the X axis, and $v(t)$ and $\omega(t)$ represent its translational and rotational velocities respectively, which are regarded as the system's control inputs. The state vector for this robot is defined by $q(t) = [x(t)\ \ y(t)\ \ \theta(t)]^T$.

When considering an implementation, the relation that exists between the system's input signals, $v(t)$ and $\omega(t)$, and the angular velocity of each wheel, $\omega_1(t)$ and $\omega_2(t)$, has been derived in Salgado (2000). Given a unicycle-type mobile robot with wheels of radius R and a distance between the wheels and the center of the vehicle of l, this relation is given by,

$$\begin{bmatrix} v(t) \\ \omega(t) \end{bmatrix} = \frac{R}{2} \begin{bmatrix} 1 & 1 \\ \frac{1}{l} & -\frac{1}{l} \end{bmatrix} \begin{bmatrix} w_1(t) \\ w_2(t) \end{bmatrix}. \tag{24}$$

As explained previously, the system is subject to an input time-delay. In the case of the unicycle-type mobile robot this means that the velocities $v(t)$ and $\omega(t)$ experience an equal time-delay. The posture kinematic model of the robot subject to an input time-delay τ is derived from (23) and is given by,

$$\dot{x}(t) = v(t-\tau)\cos\theta(t), \tag{25a}$$
$$\dot{y}(t) = v(t-\tau)\sin\theta(t), \tag{25b}$$
$$\dot{\theta}(t) = \omega(t-\tau), \tag{25c}$$

3.1.2 Omnidirectional Mobile Robot

The posture kinematic model of an omnidirectional mobile robot can be easily obtained by considering the geometric representation given in Fig. 6. The velocity components with respect to the axis $X - Y$ are obtained as in (Campion et.al, 1996) and (Canudas De Wit et.al., 1996),

$$\dot{x}(t) = u_1(t)\cos\theta(t) - u_2(t)\cos\left(-\theta(t) + \frac{\pi}{2}\right), \tag{26a}$$
$$\dot{y}(t) = u_1(t)\sin\theta(t) + u_2(t)\sin\left(-\theta(t) + \frac{\pi}{2}\right), \tag{26b}$$
$$\dot{\theta}(t) = u_3(t), \tag{26c}$$

where point $(x(t),y(t))$ is the position of the center of the robot on the plane $X - Y$ and $\theta(t)$ is the angular position with respect to the X axis. The input signals of the robot are given by $u_1(t)$, $u_2(t)$ and $u_3(t)$; where $u_3(t)$ is given as the rotational velocity of the robot, and $u_1(t)$ and $u_2(t)$ are two orthogonal vectors, of which $u_1(t)$ is aligned with the reference axis of the robot. The state vector for this robot is defined by $q(t) = [x(t)\ y(t)\ \theta(t)]^T$.

From Fig. 6 it also follows that the velocities of the wheels are related to the velocity components over the axes $X - Y$ and the rotational velocity by the transformation,

$$\begin{bmatrix} R\dot{\phi}_1(t) \\ R\dot{\phi}_2(t) \\ R\dot{\phi}_3(t) \end{bmatrix} = \begin{bmatrix} -\sin(\theta(t)+\delta) & \cos(\theta(t)+\delta) & l \\ -\sin(\theta(t)-\delta) & -\cos(\theta(t)-\delta) & l \\ \cos\theta(t) & \sin\theta(t) & l \end{bmatrix} \begin{bmatrix} \dot{x}(t) \\ \dot{y}(t) \\ \dot{\theta}(t) \end{bmatrix}, \tag{27}$$

where $\phi_i(t)$ is the angular velocity of each wheel and R is its radius, l denotes the distance between each wheel and the center of the vehicle and δ is the orientation of the wheel w.r.t. axes of the vehicle.

For a possible implementation, the relationship that exists between the input signals of the system $u_1(t)$, $u_2(t)$ and $u_3(t)$, and the angular velocity of each wheel is given by,

$$\begin{bmatrix} R\dot{\phi}_1(t) \\ R\dot{\phi}_2(t) \\ R\dot{\phi}_3(t) \end{bmatrix} = \begin{bmatrix} -\sin\delta & \cos\delta & l \\ -\sin\delta & -\cos\delta & l \\ 1 & 0 & l \end{bmatrix} \begin{bmatrix} u_1(t) \\ u_2(t) \\ u_3(t) \end{bmatrix}. \tag{28}$$

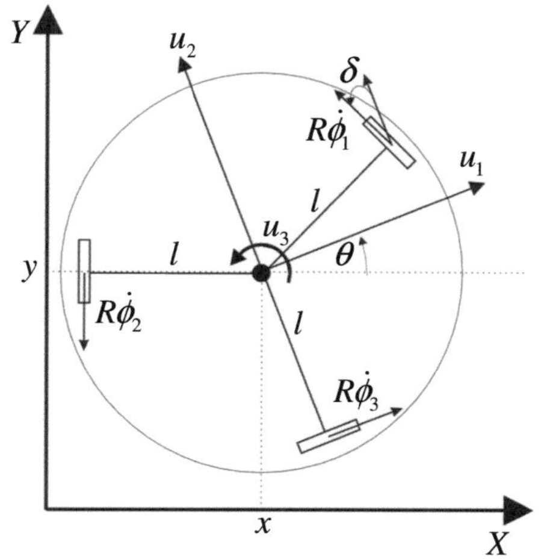

Fig. 6. Omnidirectional mobile robot.

As with the unicycle-type mobile robot, the omnidirectional mobile robot is subject to an input time-delay τ, resulting in the following posture kinematic model derived from (26),

$$\dot{x}(t) = u_1(t-\tau)\cos\theta(t) - u_2(t-\tau)\sin\theta(t) \tag{29a}$$

$$\dot{y}(t) = u_1(t-\tau)\sin\theta(t) + u_2(t-\tau)\cos\theta(t) \tag{29b}$$

$$\dot{\theta(t)} = u_3(t-\tau) \tag{29c}$$

3.2 Exact Discrete-Time Model

The discretization procedure for a nonlinear system can be found in (Kotta, 1995) and consists in obtaining the solution of the system's dynamic model along the time period corresponding to the time between two sampling instants. The class of nonlinear systems considered are,

$$\dot{x}(t) = f(x(t), u(t)). \tag{30}$$

Given a positive constant different from zero as sampling time T, the interval t_k is defined as the time interval between two sampling instants in the following way:

$$t_k = t \in [kT, kT+T), \tag{31}$$

where: $k = 0, 1, 2, 3, ...$

The general solution of the differential equation that can be proposed based on system (30) at any point of the interval t_k is given by,

$$x(t) = x(kT) + \int_{kT}^{t} f(x(\lambda), u(\lambda))d\lambda. \tag{32}$$

In the case of sampled systems, due to their digital nature, it is generally considered that the input signals of the system are modified only during the sampling instants, which means that the system's input signal $u(t)$ in (30) is constant along the interval t_k. The value of $u(t)$ will then be that which it acquired at the beginning of the interval, i.e.,

$$u(t) = u(kT). \tag{33}$$

The previous consideration allows rewriting equation (32), resulting in,

$$x(t) = x(kT) + \int_{kT}^{t} f(x(\lambda), u(kT))d\lambda. \tag{34}$$

Expression (34) represents the solution of the nonlinear system given by (30) in the time instant t within the time interval t_k. Consequently, if the solution presented in (34) is evaluated at the end of interval t_k, a nonlinear discrete-time model of the nonlinear system can be obtained as follows,

$$x((k+1)T) = x(kT) + \int_{kT}^{(k+1)T} f(x(\lambda), u(kT))d\lambda. \tag{35}$$

If the integral term in (35) has an explicit solution, then the resulting function represents an exact discrete-time model given by,

$$x((k+1)T) = x(kT) + \Phi(T, x(\lambda), u(kT)), \tag{36}$$

where:

$$\Phi(T, x(\lambda), u(kT)) = \int_{kT}^{(k+1)T} f(x(\lambda), u(kT))d\lambda. \tag{37}$$

In those cases where the integral of equation (37) can not be obtained explicitly, it is possible to obtain an approximation based on the substitution of $f(x(t), u(t))$ by its Taylor series, which results in,

$$\Phi(T, x(\lambda), x(kT), u(kT)) = \int_{kT}^{(k+1)T} \Bigg(f(x(kT), u(kT)) + (x(\lambda) - x(kT))f^{(1)}(x(kT), u(kT)) + \cdots$$
$$\cdots + \frac{(x(\lambda) - x(kT))^n f^{(n)}(x(kT), u(kT))}{n!} + \cdots \Bigg) d\lambda, \tag{38}$$

where,

$$f^{(i)}(x(kT), u(kT)) = \tfrac{\partial^i}{\partial x(kT)^i} f(x(kT), u(kT)), \qquad x \in \mathbb{R}^n, u \in \mathbb{R}^n \;. \tag{39}$$

A zero order approximation of (38) yields,

$$\Phi(T, x(kT), u(kT)) = \int_{kT}^{(k+1)T} f(x(kT), u(kT))d\lambda = Tf(x(kT), u(kT)), \tag{40}$$

which results in the following approximate discrete time model,

$$x((k+1)T) = x(kT) + Tf(x(kT), u(kT)). \tag{41}$$

3.2.1 Unicycle-Type Mobile Robot

The procedure to obtain the discrete-time model presented in this section is explained with greater detail in Orosco (2003). Consider the continuous time posture kinematic model of a unicycle-type mobile robot as given in (23). Applying the exact discretization procedure presented in (36) results in,

$$\begin{bmatrix} x((k+1)T) \\ y((k+1)T) \\ \theta((k+1)T) \end{bmatrix} = \begin{bmatrix} x(kT) \\ y(kT) \\ \theta(kT) \end{bmatrix} + \Phi(T, x(\lambda), y(\lambda), \theta(\lambda), v(kT), \omega(kT)), \tag{42}$$

where,

$$\Phi(T,x(\lambda),y(\lambda),\theta(\lambda),v(kT),\omega(kT)) = \int_{kT}^{(k+1)T} \begin{bmatrix} \cos(\theta(\lambda)) & 0 \\ \sin(\theta(\lambda)) & 0 \\ 0 & 1 \end{bmatrix} \begin{bmatrix} v(kT) \\ \omega(kT) \end{bmatrix} d\lambda. \quad (43)$$

As mentioned previously, the input signal $u(t)$ is considered to maintain a constant value $u(kT)$ along the interval t_k.

In order to obtain the exact discrete-time model, it is obvious that the instant value of angle $\theta(t)$ along the time interval t_k is required. In consequence, it is necessary to obtain the solution to the differential equation proposed for this angle in (23c). Applying equation (34) for this purpose yields,

$$\begin{aligned} \theta(t) &= \theta(kT) + \int_{kT}^{t} f(\theta(\lambda),\omega(kT)))d\lambda \\ &= \theta(kT) + [t - kT]\omega(kT). \end{aligned} \quad (44)$$

The integrals proposed in (43) are solved using the value of $\theta(t) = \theta(\lambda)$ given by (44). For the first integral this results in,

$$\begin{aligned} \int_{kT}^{(k+1)T} v(kT)\cos(\theta(\lambda))d\lambda &= \int_{kT}^{(k+1)T} v(kT)\cos(\theta(kT) + [\lambda - kT]\omega(kT))d\lambda \\ &= \frac{v(kT)}{\omega(kT)}(\sin(\theta(kT) + T\omega(kT)) - \sin\theta(kT)). \end{aligned} \quad (45)$$

For the second integral the result yields,

$$\begin{aligned} \int_{kT}^{(k+1)T} v(kT)\sin(\theta(\lambda))d\lambda &= \int_{kT}^{(k+1)T} v(kT)\sin(\theta(kT) + [\lambda - kT]\omega(kT))d\lambda \\ &= -\frac{v(kT)}{\omega(kT)}(\cos(\theta(kT) + T\omega(kT)) - \cos\theta(kT)). \end{aligned} \quad (46)$$

Finally the third integral is,

$$\begin{aligned} \int_{kT}^{(k+1)T} \omega(kT)d\lambda &= \omega(kT)\lambda\Big|_{kT}^{(k+1)T} \\ &= T\omega(kT). \end{aligned} \quad (47)$$

Applying the sum-to-product trigonometric identity on (45) and (46) results in,

$$\Phi(T,x(\lambda),y(\lambda),\theta(\lambda),v(kT),\omega(kT)) = \begin{bmatrix} 2\frac{v(kT)}{\omega(kT)}\sin\left(\frac{T\omega(kT)}{2}\right)\cos\left(\theta(kT) + \frac{T\omega(kT)}{2}\right) \\ 2\frac{v(kT)}{\omega(kT)}\sin\left(\frac{T\omega(kT)}{2}\right)\sin\left(\theta(kT) + \frac{T\omega(kT)}{2}\right) \\ T\omega(kT) \end{bmatrix}. \quad (48)$$

The exact discrete-time model of a unicycle-type mobile robot is then given by,

$$x((k+1)T) = x(kT) + 2v(kT)\frac{\sin\left(\frac{T}{2}\omega(kT)\right)}{\omega(kT)}\cos\left(\theta(kT) + \frac{T}{2}\omega(kT)\right), \tag{49a}$$

$$y((k+1)T) = y(kT) + 2v(kT)\frac{\sin\left(\frac{T}{2}\omega(kT)\right)}{\omega(kT)}\sin\left(\theta(kT) + \frac{T}{2}\omega(kT)\right), \tag{49b}$$

$$\theta((k+1)T) = \theta(kT) + T\omega(kT). \tag{49c}$$

It is worth noting that in the model, states (49a) and (49b) become undefined in the term $\frac{\sin(\frac{T}{2}\omega(kT))}{\omega(kT)}$ when $\omega(kT) = 0$. However, by l'Hôpital's rule it is possible to approximate this term by $\frac{T}{2}$. The following function is proposed to account for this situation,

$$\gamma(\omega(kT)) = \begin{cases} \frac{\sin(\frac{T}{2}\omega(kT))}{\omega(kT)} & \text{if} \quad \omega(kT) \neq 0, \\ \frac{T}{2} & \text{if} \quad \omega(kT) = 0. \end{cases} \tag{50}$$

The discrete-time exact model of a unicycle-type mobile robot is then given by,

$$x((k+1)T) = x(kT) + 2v(kT)\gamma(\omega(kT))\cos\left(\theta(kT) + \frac{T}{2}\omega(kT)\right), \tag{51a}$$

$$y((k+1)T) = y(kT) + 2v(kT)\gamma(\omega(kT))\sin\left(\theta(kT) + \frac{T}{2}\omega(kT)\right), \tag{51b}$$

$$\theta((k+1)T) = \theta(kT) + T\omega(kT). \tag{51c}$$

In the same way as (51), the exact discrete-time model of the robot with delayed inputs is obtained based on the input delayed posture kinematic model (25). Once more assuming the input signals are constant during a sampling interval, direct integration of (25c) yields,

$$\theta(t) = \theta(kT) + [t - kT]\omega(kT - \tau). \tag{52}$$

Substituting (52) in (25a) and (25b) and integrating them results in,

$$x(t) = x(kT) + \frac{v(kT-\tau)}{\omega(kT-\tau)}(\sin(\theta + T\omega(kT-\tau)) - \sin\theta), \tag{53a}$$

$$y(t) = y(kT) - \frac{v(kT-\tau)}{\omega(kT-\tau)}(\cos(\theta + T\omega(kT-\tau)) - \cos\theta), \tag{53b}$$

while the integration of (25c) in the interval $[kT, (k+1)T]$ yields,

$$\theta(t) = \theta(kT) + T\omega(kT - \tau). \tag{54}$$

After some algebraic and trigonometric manipulations the exact discrete time model of the unicycle-type mobile robot results in,

$$x((k+1)T) = x(kT) + 2v(kT-\tau)\gamma(\omega(kT-\tau))\cos\left(\theta(kT) + \frac{T\omega(kT-\tau)}{2}\right), \tag{55a}$$

$$y((k+1)T) = y(kT) + 2v(kT-\tau)\gamma(\omega(kT-\tau))\sin\left(\theta(kT) + \frac{T\omega(kT-\tau)}{2}\right), \tag{55b}$$

$$\theta((k+1)T) = \theta(kT) + T\omega(kT-\tau), \tag{55c}$$

where function $\gamma(\omega(kT-\tau))$ satisfies,

$$\gamma(\omega(kT-\tau)) = \begin{cases} \frac{\sin\left(\frac{T}{2}\omega(kT-\tau)\right)}{\omega(kT-\tau)} & \text{if} \quad \omega(kT-\tau) \neq 0, \\ \frac{T}{2} & \text{if} \quad \omega(kT-\tau) = 0. \end{cases} \tag{56}$$

For simplification purposes, the following notation will be adopted,

$$\zeta = \zeta(kT), \;\; \zeta^{\pm} = \zeta(kT \pm T), \;\; \zeta^{[\pm n]} = \zeta(kT \pm nT). \tag{57}$$

Considering the notation change proposed in (57), the exact discrete-time posture kinematic model of the unicycle-type mobile robot can be expressed as,

$$x^{+} = x + 2v^{-\tau}\gamma(\omega^{-\tau})\cos\left(\theta + \frac{T\omega^{-\tau}}{2}\right), \tag{58a}$$

$$y^{+} = y + 2v^{-\tau}\gamma(\omega^{+})\sin\left(\theta + \frac{T\omega^{-\tau}}{2}\right), \tag{58b}$$

$$\theta^{+} = \theta + T\omega^{-\tau}, \tag{58c}$$

where function $\gamma(\omega^{-\tau})$ satisfies,

$$\gamma(\omega^{-\tau}) = \begin{cases} \frac{\sin\left(\frac{T}{2}\omega^{-\tau}\right)}{\omega^{-\tau}} & \text{if} \quad \omega^{-\tau} \neq 0, \\ \frac{T}{2} & \text{if} \quad \omega^{-\tau} = 0. \end{cases} \tag{59}$$

3.2.2 Omnidirectional Mobile Robot

The exact discrete time model of the omnidirectional mobile robot subject to an input time-delay may be easily obtained by direct integration of the equations given in (29). In this sense, notice that under the assumption that the control signals are constant between sampling instances, equation (29c) produces,

$$\theta(t) = \theta(kT) + [t - kT]u_3(kT-\tau). \tag{60}$$

Substituting (60) into (29a) and (29b) and integrating as in (37) yields,

$$\begin{aligned} x(t) = x(kT) &+ \frac{u_1(kT-\tau)}{u_3(kT-\tau)}(\sin(\theta(t) + Tu_3(kT-\tau)) - \sin\theta(t)) \\ &+ \frac{u_2(kT-\tau)}{u_3(kT-\tau)}(\cos(\theta(t) + Tu_3(kT-\tau)) - \cos\theta(t)), \end{aligned} \tag{61a}$$

$$\begin{aligned} y(t) = y(kT) &- \frac{u_1(kT-\tau)}{u_3(kT-\tau)}(\cos(\theta(t) + Tu_3(kT-\tau)) - \cos\theta(t)) \\ &+ \frac{u_2(kT-\tau)}{u_3(kT-\tau)}(\sin(\theta(t) + Tu_3(kT-\tau)) - \sin\theta(t)). \end{aligned} \tag{61b}$$

After some algebraic and trigonometric manipulations the exact discrete time model of the omnidirectional mobile robot results in,

$$x^{+} = x + 2u_1^{-\tau}\gamma(u_3^{-\tau})\cos\left(\theta + \frac{Tu_3^{-\tau}}{2}\right) - 2u_2^{-\tau}\gamma(u_3^{-\tau})\sin\left(\theta + \frac{Tu_3^{-\tau}}{2}\right), \tag{62a}$$

$$y^{+} = y + 2u_1^{-\tau}\gamma(u_3^{-\tau})\sin\left(\theta + \frac{Tu_3^{-\tau}}{2}\right) + 2u_2^{-\tau}\gamma(u_3^{-\tau})\cos\left(\theta + \frac{Tu_3^{-\tau}}{2}\right), \tag{62b}$$

$$\theta^{+} = \theta_0 + u_3^{-\tau}. \tag{62c}$$

where the function $\gamma(u_3^{-\tau})$ accounts for terms that become undefined (as with the unicycle-type mobile robot) and satisfies,

$$\gamma\left(u_3^{-\tau}\right) = \begin{cases} \frac{\sin\left(\frac{T}{2}u_3^{-\tau}\right)}{u_3^{-\tau}} & \text{if } u_3^{-\tau} \neq 0, \\ \frac{T}{2} & \text{if } u_3^{-\tau} = 0. \end{cases} \tag{63}$$

3.3 Tracking Controller

Tracking a trajectory constitutes one of the simplest tasks a mobile robot can perform and constitutes the basis for achieving more complex behaviors. Designing a tracking controller based on a feedback linearization for a unicycle-type mobile robot is not straightforward due to its non-holonomic constraints, as reported by (Brockett, 1983). On the other hand, in the case of the omnidirectional mobile robot, a tracking controller by means of a full state feedback is possible, constituting the best option for this type of robot.

3.3.1 Unicycle-Type Mobile Robot

According to Lefeber et.al (2001), the problem of path-tracking for a unicycle-type mobile robot can be stated as the requirement for the robot to follow a reference trajectory with state $q_r(t) = [x_r(t)\ \ y_r(t)\ \ \theta_r(t)]^T$ generated by an exosystem with kinematics,

$$\dot{x}_r(t) = v_r(t)\cos\theta_r(t), \tag{64a}$$

$$\dot{y}_r(t) = v_r(t)\sin\theta_r(t), \tag{64b}$$

$$\dot{\theta}_r(t) = \omega_r(t), \tag{64c}$$

where $v_r(t)$ and $\omega_r(t)$ are continuous functions of time given as the reference velocities by,

$$v_r(t) = \sqrt{\dot{x}_r^2(t) + \dot{y}_r^2(t)}, \tag{65a}$$

$$\omega_r(t) = \frac{\dot{x}_r(t)\ddot{y}_r(t) - \ddot{x}_r(t)\dot{y}_r(t)}{\dot{x}_r^2(t) + \dot{y}_r^2(t)}. \tag{65b}$$

From Fig. 5, it follows that the position errors between the mobile robot and the reference system can be expressed in terms of the error coordinates $q_e(t) = [x_e(t)\ \ y_e(t)\ \ \theta_e(t)]^T$ on a moving coordinate frame mounted on the robot, i.e.

$$\begin{bmatrix} x_e(t) \\ y_e(t) \\ \theta_e(t) \end{bmatrix} = \begin{bmatrix} \cos\theta(t) & \sin\theta(t) & 0 \\ -\sin\theta(t) & \cos\theta(t) & 0 \\ 0 & 0 & 1 \end{bmatrix} \begin{bmatrix} x_r(t) - x(t) \\ y_r(t) - y(t) \\ \theta_r(t) - \theta(t) \end{bmatrix}, \tag{66}$$

which yield the following error dynamics when derived w.r.t to time,

$$\dot{x}_e(t) = \omega(t)y_e(t) - v(t) + v_r(t)\cos\theta_e(t), \tag{67a}$$
$$\dot{y}_e(t) = -\omega(t)x_e(t) + v_r(t)\sin\theta_e(t), \tag{67b}$$
$$\dot{\theta}_e(t) = \omega_r(t) - \omega(t). \tag{67c}$$

Since the error dynamics are also affected by the time-delay they are actually expressed as,

$$\dot{x}_e(t) = \omega(t-\tau)y_e(t) - v(t-\tau) + v_r(t)\cos\theta_e(t), \tag{68a}$$
$$\dot{y}_e(t) = -\omega(t-\tau)x_e(t) + v_r(t)\sin\theta_e(t), \tag{68b}$$
$$\dot{\theta}_e(t) = \omega_r(t) - \omega(t-\tau). \tag{68c}$$

Due to the fact that the delayed error dynamics (68) can not be integrated explicitly, their approximate Euler-discretization is computed, and results in,

$$x_e^+ = x_e + T(\omega^{-\tau}y_e - v^{-\tau} + v_r\cos\theta_e), \tag{69a}$$
$$y_e^+ = y_e + T(-\omega^{-\tau}x_e + v_r\sin\theta_e), \tag{69b}$$
$$\theta_e^+ = \theta_e + T(\omega_r - \omega^{-\tau}). \tag{69c}$$

In the continuous time case, the approach presented in (Lefeber et.al, 2001) proposes the use of a cascaded structure based on the error dynamics (67) that results in the use of linear control laws. The same idea has been used in (Nešić and Loría, 2004) for the discrete time case, in which the following controller has been proposed,

$$\omega^{-\tau} = \omega_r + c_1\theta_e, \tag{70a}$$
$$v^{-\tau} = v_r + c_2x_e + T\vartheta, \tag{70b}$$

where c_1, c_2, v_r and ω_r are the same as in the continuous-time control law proposed in Lefeber et.al (2001). The term $T\vartheta$ is an extra control input should be designed to improve the performance of the system.

Considering a correcting term given by $\vartheta = -\frac{c_3\omega_r y_e}{T}$, where c_3 is a gain, and modifying (70a) in order to accommodate more aggressive angle variations, the following controller is proposed,

$$\omega^{-\tau} = \omega_r + c_1\sin\theta_e, \tag{71a}$$
$$v^{-\tau} = v_r + c_2x_e - c_3\omega_r y_e. \tag{71b}$$

It is worth noting that a practical feedback should be synthesized at time t or, more precisely, at time instant kT. In this case, feedback (71) results in an anticipative controller given by,

$$\omega = \omega_r^{+\tau} + c_1\sin\theta_e^{+\tau}, \tag{72a}$$
$$v = v_r^{+\tau} + c_2x_e^{+\tau} - c_3\omega_r^{+\tau}y_e^{+\tau}, \tag{72b}$$

which yields the anticipated error coordinates,

$$\begin{bmatrix} x_e^{+\tau} \\ y_e^{+\tau} \\ \theta_e^{+\tau} \end{bmatrix} = \begin{bmatrix} \cos\theta^{+\tau} & \sin\theta^{+\tau} & 0 \\ -\sin\theta^{+\tau} & \cos\theta^{+\tau} & 0 \\ 0 & 0 & 1 \end{bmatrix} \begin{bmatrix} x_r^{+\tau} - x^{+\tau} \\ y_r^{+\tau} - y^{+\tau} \\ \theta_r^{+\tau} - \theta^{+\tau} \end{bmatrix}. \tag{73}$$

Remark 3.1 *The control law proposed in (72) with the error coordinates (73) are non-causal expressions which require the state values τ samples of time into the future. The Smith-like prediction strategy proposed in Section 2 will be used to generate these values.*

3.3.2 Omnidirectional Mobile Robot

Given the discrete-time input delayed posture kinematic model of the omnidirectional robot (62), the following output is proposed in order to obtain a fully linearizing feedback controller,

$$y = h(q) = \begin{bmatrix} h_1(q) \\ h_2(q) \\ h_3(q) \end{bmatrix} = \begin{bmatrix} x \\ y \\ \theta \end{bmatrix}. \tag{74}$$

Since the relative degree of system (62) w.r.t the proposed inputs in (74) is 3, a fully linearizing feedback controller may be implemented. Deriving the output once yields,

$$\begin{bmatrix} h_1^+(q,u^{-\tau}) \\ h_2^+(q,u^{-\tau}) \\ h_3^+(q,u^{-\tau}) \end{bmatrix} = \begin{bmatrix} x + 2u_1^{-\tau}\gamma(u_3^{-\tau})\cos\left(\theta + \frac{Tu_3^{-\tau}}{2}\right) - 2u_2^{-\tau}\gamma(u_3^{-\tau})\sin\left(\theta + \frac{Tu_3^{-\tau}}{2}\right) \\ y + 2u_1^{-\tau}\gamma(u_3^{-\tau})\sin\left(\theta + \frac{Tu_3^{-\tau}}{2}\right) + 2u_2^{-\tau}\gamma(u_3^{-\tau})\cos\left(\theta + \frac{Tu_3^{-\tau}}{2}\right) \\ \theta_0 + Tu_3^{-\tau} \end{bmatrix}. \tag{75}$$

In order to obtain a controller that fully decouples the system, a set of auxiliary controllers v_1, v_2 y v_3 are proposed such that the closed-loop system satisfies,

$$\begin{bmatrix} h_1^+(q,u^{-\tau}) \\ h_2^+(q,u^{-\tau}) \\ h_3^+(q,u^{-\tau}) \end{bmatrix} = \begin{bmatrix} v_1 \\ v_2 \\ v_3 \end{bmatrix}. \tag{76}$$

The trajectory tracking problem is solved by proposing,

$$v_1 = x_r^+ - k_1 e_1, \tag{77a}$$

$$v_2 = y_r^+ - k_2 e_2, \tag{77b}$$

$$v_3 = \theta_r^+ - k_3 e_3, \tag{77c}$$

with:

$$e_x = x - x_r, \tag{78a}$$

$$e_y = y - y_r, \tag{78b}$$

$$e_\theta = \theta - \theta_r, \tag{78c}$$

where $q_r = [x_r \;\; y_r \;\; \theta_r]^T$ constitute the robot's desired trajectories, $q_e = [e_x \;\; e_y \;\; e_\theta]^T$ denote the tracking positions errors and $K = [k_1 \;\; k_2 \;\; k_3]^T$ are the gains of the auxiliary controllers.

Using (75) and (76), a fully linearizing controller is given by,

$$\begin{bmatrix} u_1^{-\tau} \\ u_2^{-\tau} \\ u_3^{-\tau} \end{bmatrix} = \begin{bmatrix} \dfrac{[v_1 - x]\cos\left(\theta + \frac{Tu_3^{-\tau}}{2}\right) + [v_2 - y]\sin\left(\theta + \frac{Tu_3^{-\tau}}{2}\right)}{2\gamma(u_3^{-\tau})} \\ \dfrac{[v_2 - y]\cos\left(\theta + \frac{Tu_3^{-\tau}}{2}\right) - [v_1 - x]\sin\left(\theta + \frac{Tu_3^{-\tau}}{2}\right)}{2\gamma(u_3^{-\tau})} \\ \dfrac{v_3 - \theta}{T} \end{bmatrix}. \tag{79}$$

It is possible to obtain the controllers u_1, u_2 and u_3 in the current time instant by shifting expression (79) τ units of time, i.e.,

$$\begin{bmatrix} u_1 \\ u_2 \\ u_3 \end{bmatrix} = \begin{bmatrix} \dfrac{\left[v_1^\tau - x^\tau\right]\cos\left(\theta^\tau + \frac{Tu_3}{2}\right) + \left[v_2^\tau - y^\tau\right]\sin\left(\theta^\tau + \frac{Tu_3}{2}\right)}{2\gamma(u_3)} \\ \dfrac{\left[v_2^\tau - y^\tau\right]\cos\left(\theta^\tau + \frac{Tu_3}{2}\right) - \left[v_1^\tau - x^\tau\right]\sin\left(\theta^\tau + \frac{Tu_3}{2}\right)}{2\gamma(u_3)} \\ \dfrac{v_3^\tau - \theta^\tau}{T} \end{bmatrix}, \tag{80}$$

where,

$$v_1^\tau = x_r^{\tau+1} - k_1 e_1^\tau, \tag{81a}$$
$$v_2^\tau = y_r^{\tau+1} - k_2 e_2^\tau, \tag{81b}$$
$$v_3^\tau = \theta_r^{\tau+1} - k_3 e_3^\tau, \tag{81c}$$

with:

$$e_1^\tau = x^\tau - x_r^\tau, \tag{82a}$$
$$e_2^\tau = y^\tau - y_r^\tau, \tag{82b}$$
$$e_3^\tau = \theta^\tau - \theta_r^\tau. \tag{82c}$$

Remark 3.2 *Equivalent to Remark 3.1, the control law proposed in (80)-(82) is non-causal and will therefore require a prediction strategy such as the one presented in Section 2 in order to obtain the state values τ samples of time into the future.*

Remark 3.3 *Note that the control law for the unicycle-type mobile robot (72) with the error coordinates (73) and for the omnidirectional robot (80)-(82) both require the values of the desired trajectory τ and $\tau+1$ samples of time into the future. If these values could be provided, the mobile robot would track the desired trajectories in the current time instant, i.e. $q_r = [x_r \;\; y_r \;\; \theta_r]^T$. Although ideal, knowing the reference trajectory a priori is not possible in most cases. Consequently this constraint can be relaxed by making use of the desired trajectories in the current time instant kT and $(k+1)T$. The result will be that the mobile robot will track a delayed version of the desired trajectories, i.e., $q_r^{-\tau} = [x_r^{-\tau} \;\; y_r^{-\tau} \;\; \theta_r^{-\tau}]^T$. Simulation results in Section 5 will further clarify this point.*

4. Time-Delay Compensation

The prediction strategy and noncausal control laws of the previous sections are seamlessly integrated to compensate for an input time-delay in a mobile robot. Moreover, a simple extension to the nonlinear predictor's structure is proposed in order to cope with bilateral time-delays, moving towards implementing a fully telecontrolled system.

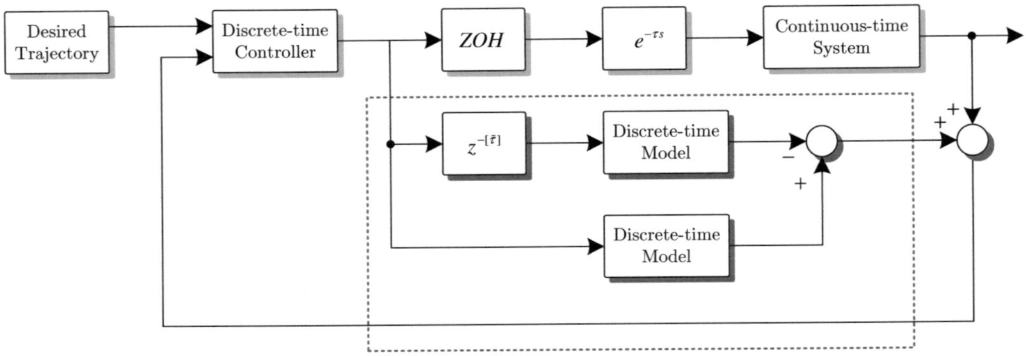

Fig. 7. Input time-delay compensation.

4.1 Input Time-Delay Compensation

The discrete-time nonlinear Smith-like predictor presented in Section 2 can be used to implement the noncausal control laws obtained in Section 3 for either the unicycle-type or the omnidirectional mobile robot in order to compensate the presence of an input time-delay. The integration of both schemes is straightforward and shown in Fig. 7.

The block denoted as *Reference Trajectory* provides the reference signals q_r, v_r and ω_r for the mobile robots. As noted in Remark 3.3, the most general assumption is that the reference trajectory will be provided w.r.t to the current sampling instant and therefore the mobile robot will track a delayed version of this signal. The noncausal control law is computed in the *Discrete Control* block and requires as input the predicted states of the system and the reference trajectory. The Smith-like predictor is implemented in the blocks within the dotted line and contains the time-delay's model and the delayed and delay free models of the system. The blocks representing the physical robot and the time-delay affecting its input are also shown in the figure.

4.2 Bilateral Time-Delay Compensation

When studying the behavior of the predictor, the controller and the mobile robot for the input time-delay case, several observations can be made. Due to the structure of the predictor, one might say that it acts as a kind of full state observer, where it looks to reproduce the system's behavior without a time-delay. Due to the fact that the controller receives the predictor's output, which is not experiencing an input time-delay, it is able to compute a control law which will track the reference trajectory in the current time instant. It then follows that, if the control signal is delayed, the system receiving it will track a delayed version of the reference trajectory.

When considering a bidirectional time-delay, a delay in the system's outputs is also present. In this case the system is subject to a forward τ_f and backward τ_b time-delay, as explained in (Hokayem and Spong, 2006). If the Smith-like predictor as presented previously is applied to this case, the robot's performance is obviously degraded. The reason for this is, of course, the output time-delay affecting the robot. What happens is that the comparison carried out between the delayed outputs of the system and the delayed model is no longer relevant due to the fact that the model is not accounting for the output time-delay. As this happens, the predictor starts feeding the controller with information that is not relevant, which in turn, produces control outputs that are not adequate. This inadequate outputs are then received by the system after a time-delay and a "vicious" cycle is started.

It appears that the problem in the bidirectional delay case when considering the Smith-like predictor is that the correcting term in the predictor is not bringing the predictor's states closer to the system's states. If this term was able to draw the systems closer, then the predictor would be able to start providing the controller with the correct information and thus, the robot's performance would improve.

Assume for the time being that the forward and backward time-delays are equal and constant, i.e. $\tau_b = \tau_f = \tau$. Assume also that the time-delays in the communication channel are modeled perfectly, i.e. $\tilde{\tau} = \tau$. Note how the backward time-delay does not affect the robot's states, yielding the usual input delayed continuous time posture kinematic model for the unicycle-type (25) and omnidirectional (29) mobile robots.

Nevertheless, once the robot's outputs travel back to the controller side they are affected by the backward time-delay, resulting in,

$$q(t-\tau) = \begin{bmatrix} x(t-\tau) \\ y(t-\tau) \\ \theta(t-\tau) \end{bmatrix}. \tag{83}$$

Considering (83), a simple but logical solution to produce a relevant comparison within the predictor would be to also delay the outputs of the delayed model in the predictor by $\tau_b = \tau$. The modified continuous time nonlinear predictor may be characterized by the following expressions,

$$\dot{\tilde{x}}(t) = \tilde{f}(\tilde{x}(t)) + \tilde{g}(\tilde{x}(t))u(t), \tag{84a}$$

$$\dot{\hat{x}}(t) = \tilde{f}(\hat{x}(t)) + \tilde{g}(\hat{x}(t))u(t-\tau), \tag{84b}$$

$$\delta x(t) = \tilde{x}(t) - \hat{x}(t-\tau), \tag{84c}$$

whereas the discrete time version is given by,

$$\tilde{x}(k+1) = \tilde{f}(\tilde{x}(k)) + \tilde{g}(\tilde{x}(k)u(k)), \tag{85a}$$

$$\hat{x}(k+1) = \tilde{f}(\hat{x}(k)) + \tilde{g}(\hat{x}(k)u(k-\tau)), \tag{85b}$$

$$\delta x(k) = \tilde{x}(k) - \hat{x}(k-\tau). \tag{85c}$$

Considering (84) and (85), the term entering the controller will be given by $x^*(t) = \tilde{x}(t)$ in the continuous time case and $x^*(k) = \tilde{x}(k)$ in the discrete time case respectively. Given the new correcting term the comparison carried out within the predictor is once more able to bring the predictor's states closer to the system's states and thus, the controller is fed with the "correct" outputs by the predictor. This would mean in turn that the robot will be provided by the controller with the adequate input signals to track a delayed version of the reference signal. The block diagram of the modified bilateral predictor is depicted in Figure 8.

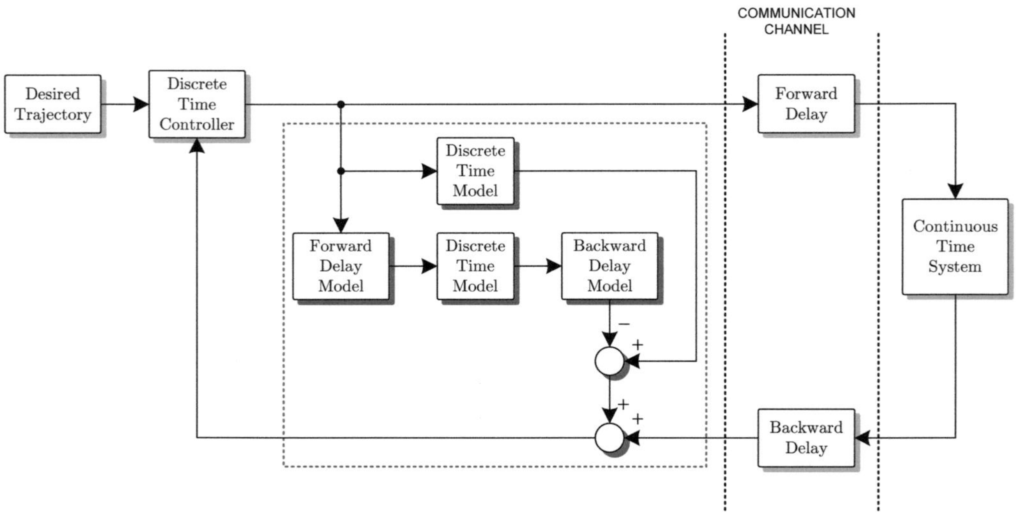

Fig. 8. Bilateral prediction strategy.

5. Simulation Results

This section includes numerous simulations in order to illustrate the performance of the bilateral time-delay compensation control scheme proposed in Section 4.

5.1 Unicycle-type Mobile Robot

For the unicycle-type mobile robot, in the first simulation the reference trajectory is a lemniscate with half length and width of 2m, centered at [0,0]m and with a desired tracking angular velocity of 0.2m/s. The reference trajectory in the second simulation is given by a sinusoid originating at $[0.5, 0.5]m$, with an amplitude of 1.5m, an angular frequency of 0.5rad/s, a linear velocity factor of 0.1m/s and oriented at $\pi/4$rad. The initial condition for the mobile robot is $q(0) = [0.1 \;\; 0.1 \;\; 0.0]^T$ in the first simulation and $q(0) = [-0.3 \;\; 1.0 \;\; 0.0]^T$ in the second simulation. The system's models are assumed to be perfect and are initialized with $\tilde{q}(0) = \hat{q}(0) = [0.0 \;\; 0.0 \;\; 0.0]^T$ in both simulations. The controller gains in both cases are set to $c_1 = c_2 = c_3 = 3.0$, and the forward and backward time-delays induced by the communication channel, which are assumed to be perfectly modeled, are given by $\tau_f = \tau_b = 0.25$sec. Both simulations are sampled every $T = 0.05$sec (20Hz) and are carried out during 50sec.

The behavior of the mobile robot in the $X - Y$ plane (red) and its desired trajectory (blue) are depicted in Fig. 9 for both simulations. The small cross markers denote the simulations' initial state and the circular markers their intermediate (25sec) and final state (50sec). Due to the time-delay, the mobile robot lags its desired trajectory. The lag distance is determined by the time-delay's magnitude and the characteristics of the reference trajectory.

The first and third columns of Fig. 10 depict the mobile robot's states (red) and their reference (blue) for both simulations, while their errors are shown in the second and fourth columns of the same figure. The errors are computed w.r.t to a version of the reference trajectory that is delayed by τ, so that both the states and the reference are in the same time base. From the

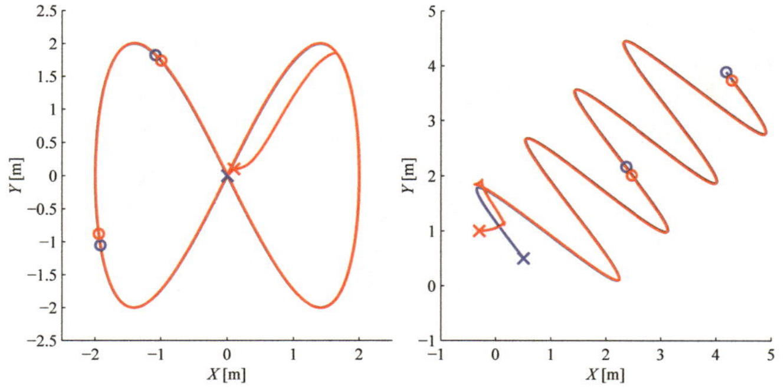

Fig. 9. Unicycle-type mobile robot: behavior in the $X - Y$ workspace.

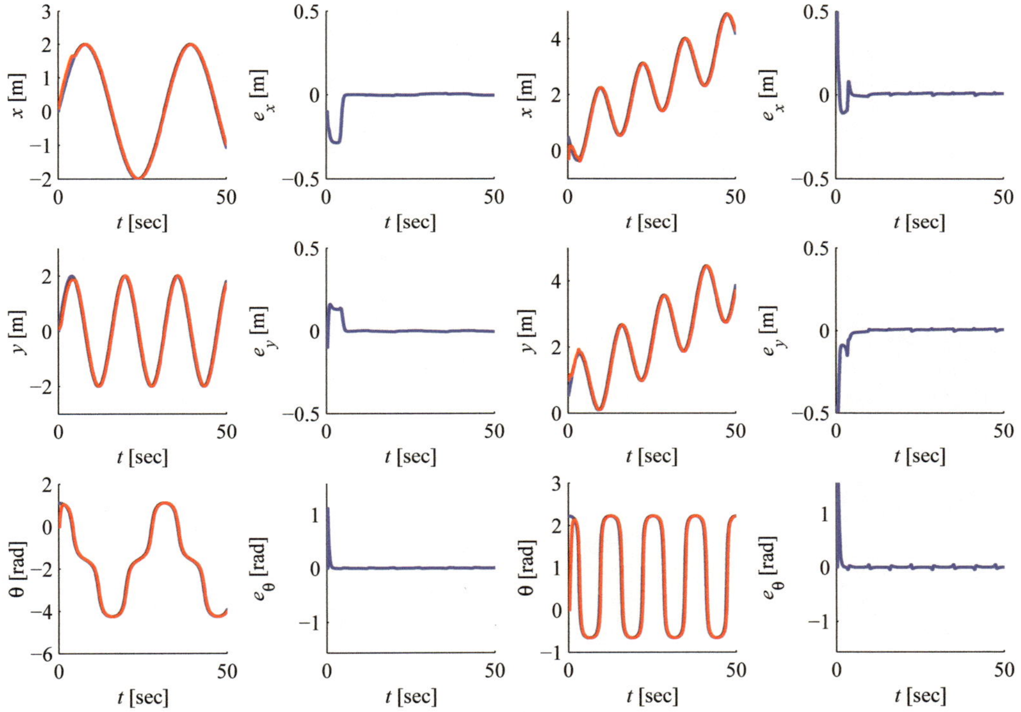

Fig. 10. Unicycle-type mobile robot: states and errors.

state plots a slight displacement between the reference trajectory and the states can be noticed due to the time-delay. This displacement is not very clear since the simulation time is too large compared with the time-delay. Given the fact that the error plots converge or practically converge to zero, it can be concluded that the unicycle-type mobile robot tracks its reference trajectory after a time τ.

5.2 Omnidirectional Mobile Robot

The reference trajectories and simulation conditions for the omnidirectional mobile robot are exactly the same as for the unicycle-type mobile robot, with the exception that the desired orientation is provided independently at a fixed value of $\pi/2$rad. This is due to the fact that this type of robot does not have non-holonomic constraints and thus can follow any trajectory with an arbitrary orientation. The initial condition for the mobile robot is $q(0) = [1.0\ \ 1.0\ \ 0.0]^T$ in the first simulation and $q(0) = [2.0\ \ -2.0\ \ 0.0]^T$ in the second simulation. The controller gains in both cases are set to $k_1 = -0.85$, $k_2 = -0.85$, $k_3 = 0.3$.

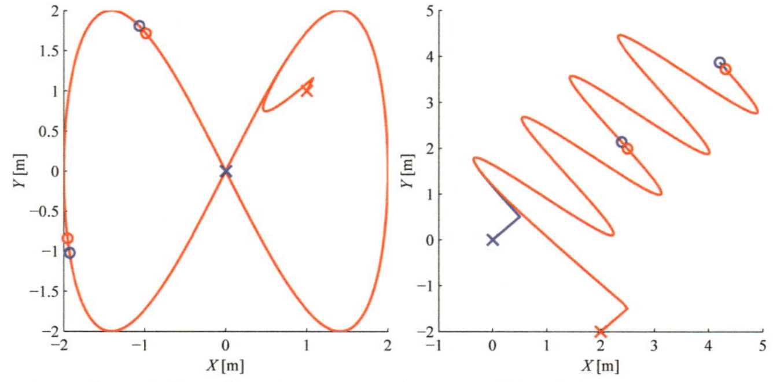

Fig. 11. Omnidirectional mobile robot in its workspace $X - Y$.

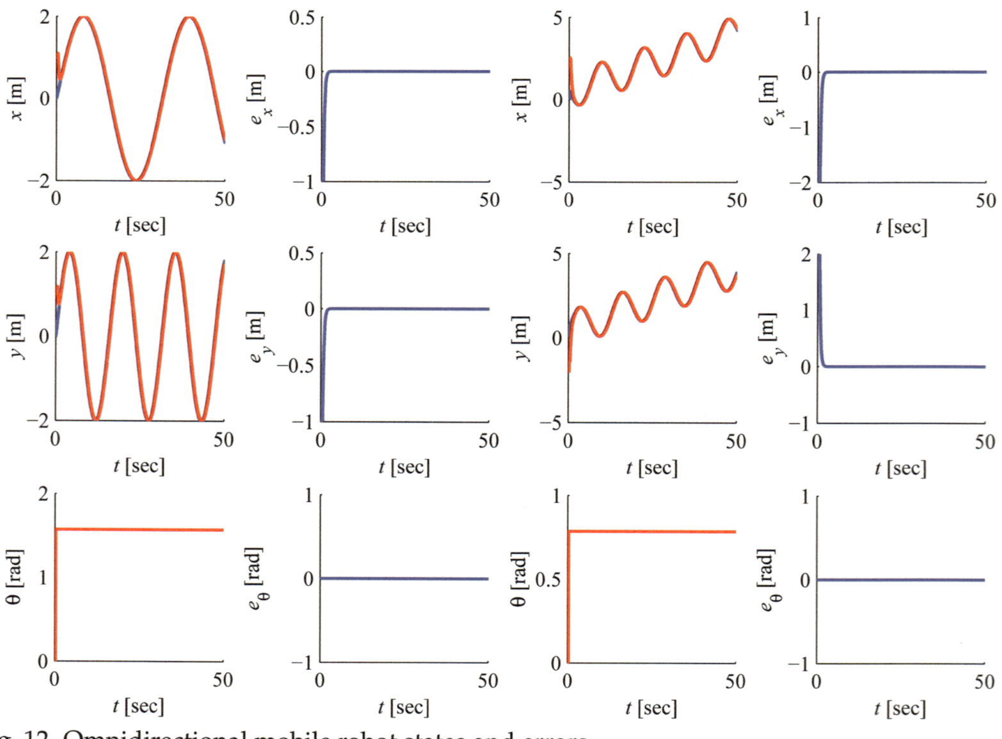

Fig. 12. Omnidirectional mobile robot states and errors.

As with the unicycle-type mobile robot, the behavior of the omnidirectional mobile robot in the $X - Y$ plane (red) and its desired trajectory (blue) is depicted in Fig. 11 for both simulations. The small cross and circular markers denote the same elements as in Fig. 9. Once more, the mobile robot lags its desired trajectory due to the time-delay.

The mobile robot's states (red) compared to their reference (blue) together with their errors are shown in the different columns of Fig. 12 using the same distribution as in Fig. 10. Once more the state plots show a slight displacement between the reference trajectory and the states which is due to the time-delay and is barely noticeable since the simulation time to time-delay ratio is very high. The error plots once more converge or practically converge to zero, which means that the omnidirectional mobile robot is tracking its reference trajectory after a time τ.

Remark 5.1 *The proposed prediction structure has shown to be highly sensitive to a mismatch in the orientation's initial condition. This issue appears in both types of mobile robots and is the reason why the initial orientation in all the simulations was set to zero. In other words, if the models' and system's initial orientations differ, the system's performance is affected. Although not ideal, in a practical setting the robot's initial orientation can be determined beforehand in order to set the models' initial condition.*

6. Conclusions and Future Work

A Smith-type predictor compensator has been proposed for two types of wheeled mobile robots subject to either an input or a bilateral time-delay. Using the exact discrete-time posture kinematic model of the mobile robots, the predictor's structure allows to implement noncausal control laws whose design is based on the delay free system. Numerical simulations show that the mobile robots will track a delayed version of the reference trajectory.

Many possible improvements to the proposed prediction-control scheme are possible. As noted by (Michiels and Niculescu, 2007), most of the work regarding the Smith predictor focuses on its robustness and its disturbance rejection characteristics. These issues are definitely relevant in the context of mobile robotics and should therefore be followed closely. For example, an adaptive algorithm that compensates for mismatch in the time-delay model would significantly improve the system's performance. A further extension of this work to robotic manipulators is also a possibility. In this case, issues such as the reflection of the contact and the driving forces has to be considered.

7. References

Brockett, R.W. (1983). Asymptotic stability and feedback stabilization. In *Differential Geometric Control Theory, R.W. Brockett, R.S. Milman and H.J. Sussmann, Eds.*, Birkhäuser, Boston, pp. 181-191.

Canudas De Wit, C., Siciliano, B., Bastin, G., Brogliato, B., Campion, G., DAndrea-Novel, B., De Luca, A., Khalil, W., Lozano,R., Ortega, R., Samson, C. and Tomei, P. (1996). *Theory of Robot Control*. Springer-Verlag, London.

Campion, G. , Bastin, G. and DAndrea-Novel, B. (1996). Structural properties and classification of kinematic and dynamic models of wheeled mobile robots. *IEEE Transactions on Robotics and Automation*, 12(1), pp. 4762.

Henson, M. A. and Seborg, D. E.. (1994). Time delay compensation for nonlinear processes. *Industrial Engineering Chem.*, 33(6):14931500.

Hokayem, P.F. and Spong, M.W. (2006). Bilateral Teleoperation: An Historial Survey. *Automatica*, 42(12):2035-2057.
Jerome, N.F. and Ray, W.H. (1986). High-performance multivariable control strategies for systems having time delays. *AIChE Journal*, 32(6):914931.
Kotta, U. (1995). *Inversion method in the discrete-time nonlinear control systems synthesis problems*. Springer Berlin/Heidelberg, Berlin, Germany.
Kravaris, C. and Chung, C.B. (1987). Nonlinear feedback synthesis by global input/output linearization. *AIChE Journal*, 33(4):592-603.
Kravaris, C. (1988). Input/Output linearization: A nonlinear analog of placing poles at process zeros. *AIChE Journal*, 34(11):1803-1812.
Kravaris, C. and Wright, R. A. (1989). Deadtime compensation for nonlinear processes. *AIChE Journal*, 35(9):15351542.
Lefeber, E., Jakubiak, J., Tchoń, K. and Nijmeijer, H. (2001). Observer based kinematic tracking controllers for a unicycle-type mobile robot. In *Proceedings of the 2001 IEEE International Conference on Robotics & Automation*, Seoul, Korea.
Michiels, W. and Niculescu, S.I. (2007). *Stability and Stabilization of Time-Delayt Systems: An Eigenvalue Based Approach*. SIAM, Philadelphia.
Morari, M. and Zafiriou, E. (1989). *Robust process control*. Prentice Hall, New Jersey.
Nešić, D. and Loría, A. (2004). On uniform asymptotic stability of time-varying parameterized discrete-time cascades. *IEEE Transactions on Automatic Control*, 49(6):875-887.
Orosco-Guerrero, R., Velasco-Villa, M. and Aranda-Bricaire, E. (2004). Discrete-time controller for a wheeled mobile robot. In *Proc. XI Latin-American Congress of Automatic Control*, La Habana, Cuba.
Salgado-Jimnez, T. (2000). *Design, implementation and control of a mobile robot*. Masther's thesis, CINVESTAV IPN, Mexico.
Smith, O.J.M. (1957). Closer control of loops with dead time. *Chem. Eng. Prog.*, 53, pp. 217-219.

4

Stereo Vision System for Remotely Operated Robots

Angelos Amanatiadis and Antonios Gasteratos
Democritus University of Thrace
Greece

1. Introduction

Remotely operated robots, functioning in hazardous and time critical environments, have significant requirements for control and visual information (Davids (2002), Murphy (2004)). The control systems are supposed to guarantee a precise timely response in order to prevent fatal scenarios in bomb disposal operations or in life rescue missions. Significant role to these operating scenarios play the concurrent visual information provided to the remote operators by the on-board mounted cameras.

Visual information (Fong & Thorpe (2001), Desouza & Kak (2002)) is often displayed in one or more monitors depending on the number of on-board mounted cameras. In sophisticated and multi-tasking robots more than one operators are performing certain actions. Especially, in case of robots with grippers and robotic arms, one operator might be dedicated only with the maneuvering and controlling of the robotic arms or grippers. In these operation scenarios, the dedicated user must be focused only on this task and furthermore should have the best visual understanding of the working field. The widely used equipment and gear for these assignments is a Head Mounted Display (HMD) and an attached head tracker.

The HMD projects visual feedback of the remote robot in front of operator eyes. A single camera feedback projection in both eyes is not so significant since the result in operator's perception is the same as being watched from a single monitor. Thus, a pair of cameras are used instead, in order to provide a real stereo feedback to the operator's HMD, thus enhancing his visual perception and improving the sense of depth (Willemsen et al. (2004)). Consequently, operators can judge situations and perform actions more efficiently based on the qualitative information of the synchronized stereo video streams. The use of a head tracker expands the operator's functions while it offers a hands-free ability to remotely control the pose of the robot head. The inertial measurement devices used for the head tracking usually contain rate gyroscopes (gyros) and accelerometers. The measurements of the inertial sensor can be processed and transmitted as control signals to the remote robot.

Many different interfaces have been presented in literature recently. A method of robot teleoperation that allows a human operator to control a robot manipulator is presented in (Kofman et al. (2005)). It uses a non-contacting, vision-based, human-robot interface for both the communication of the human motion to the robot and for feedback of the robot motion to the human operator. However, this visual feedback does not give the operator the depth visual information, which is necessary for this critical task. In (Bluethmann et al. (2003)), a sophisticated anthropomorphic robot is developed for space operations. It is comprised of a stereo

Fig. 1. The Stereo vision system.

head which transmits the video feedback to the operator through a HMD. The same human-machine interface is developed also in (Tachi et al. (2003)) for robot control. Both implementations however, require sophisticated and expensive equipment and are built with proprietary software. In (Marin et al. (2005)), an on-line robot architecture is presented. It enables the control of a robot by interacting with an advanced user interface with very promising results but the real-time constraint for control can not be guaranteed.
In this chapter, we focus on a human-machine interface which guarantees the real-time control of a binocular robotic head. Furthermore, a stereo video streaming transmission with low latency is presented. This hands-free interface is implemented exclusively with open source software on a Linux-based Real-time Operating System. The control and video architecture satisfies also the demands of recent sophisticated telerobotics for flexibility and expandability.

2. System Design

The functions of the stereo vision system, as shown in Fig. 1, are separated into video streaming and motion control and are both implemented with the use of two host computers. The first host computer is called Mobile Mechatronic Unit (MMU), and is placed on the mobile platform where the stereo vision head is operating. The second host computer comprises the Mobile Control Unit (MCU) which is placed on the remote control center. There, the remote operator wears the HMD with the attached head tracker, as shown in Fig. 2.
The video streaming presents high computational burden and resource demand while it requires the full usage of certain instruction sets of a modern microprocessor. In contrast, motion control includes filtering and signal processing from the head tracker output and requires the operation system to be able to execute real-time tasks. This demand for high multimedia performance and real-time motion control is realized by a computer structure consisting of two high performance computers with RT-Linux operating system for the MMU and MCU host computers. However, the two main tasks of the video streaming and the motion control will be performed in different kernel layers due to their different requirements. Furthermore, RT-

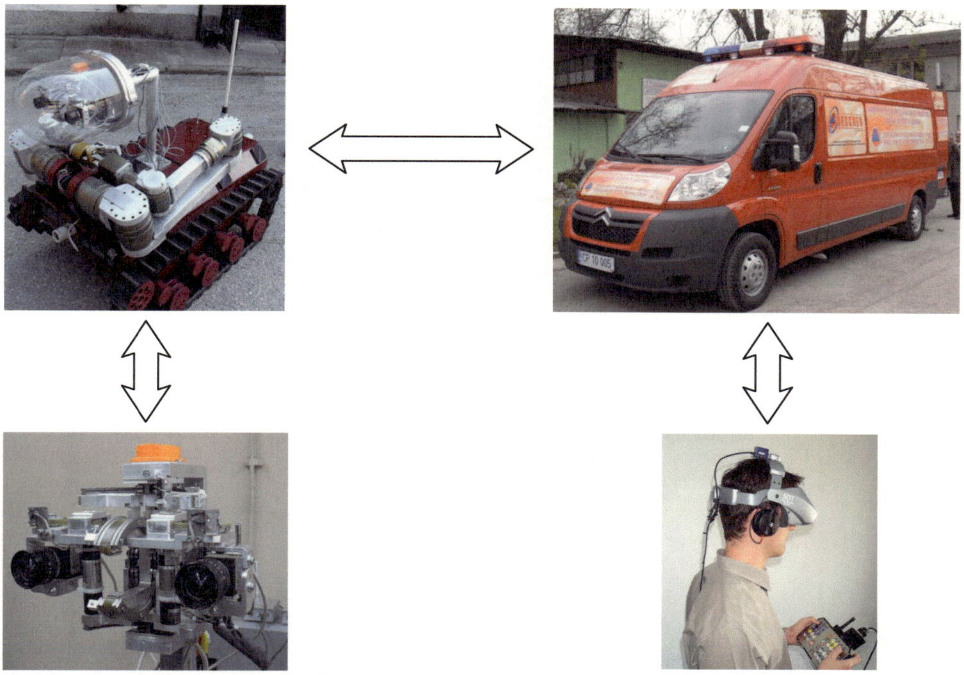

Fig. 2. The remotely operated robot.

Linux operating system was chosen in order to ensure that the different operation and communications loops occur at deterministic rates, and that safety critical tasks are performed reliably. The two host computers are connected together with a wireless high speed network. The communication protocol between the computers uses a higher level abstraction, built on top of sockets, meeting the requirements for low latency and priority handling. Apart from the libraries, the communication protocol consists of a server daemon residing on each side MCU and MMU and acts as gateway to the other side. Each server daemon is in charge of delivering the messages received by the remote end to the clients in its network, and forwarding the messages received by clients in its network, to the remote end, through the wireless link. Furthermore, there is a priority list that refers to the priority assigned to each operation. In the MCU and MMU communication subsystem, the priority handling is realized by means of the cooperation between the server protocol and quality of service features such as packet identification rules, service flow classes for bandwidth reservation, latency and jitter control on the identified packets, and quality of service classes with associated service flow classes. Figure 3 shows the flowchart of the stereo vision software architecture. The right part shows the control flow from the head tracker starting from the MCU and ending to the motors of the stereo head. The left flow shows the video stream from the stereo head cameras of the MMU to the player in the MCU.

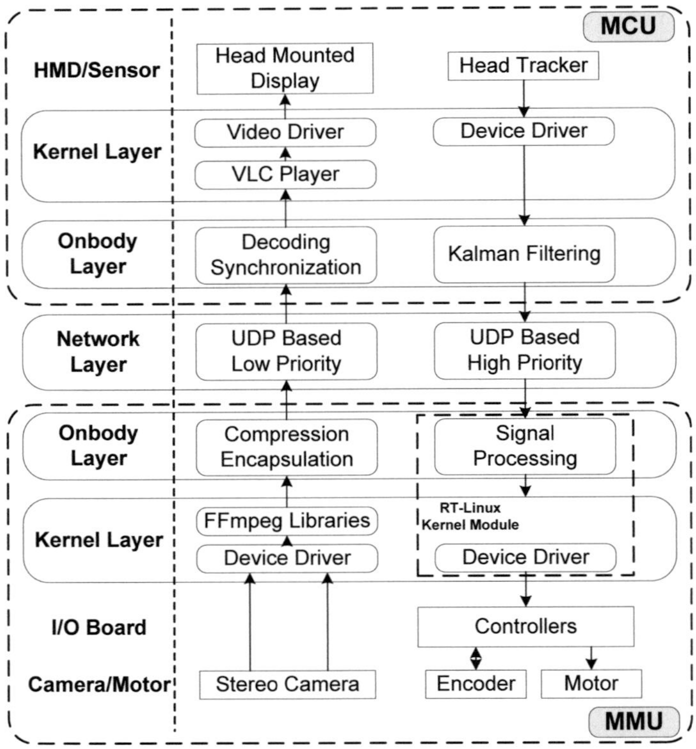

Fig. 3. Flowchart of the stereo vision software architecture.

3. Control System

3.1 Hardware Architecture

A head tracker inertial measurement unit was used to obtain high update rate measurements. Its internal low-power signal processor provides 3D orientation as well as kinematic data of 3D acceleration and 3D rate of turn (rate gyro). The data used for the head tracking is the pitch and yaw in order to send the pan and tilt commands to the teleoperated stereo head, respectively. The chosen interface used for connecting the sensor to the MCU computer is the RS-232, in order to have full access to the basic level of the sensor unit and a full compatibility for the drivers, since no serial-to-USB converter drivers are needed. The second sensor attached on the stereo head mechanism, as shown in Fig. 1, is used for the stabilization of the stereo head (Amanatiadis et al. (2007)). Special attention was paid for the placement of this inertial sensor. Possible errors and distortions from the strong currents of the servo motors can be quite large enough in order to deteriorate the inertial measurements (Roetenberg et al. (2005)). A global reset is performed each time the HMD sensor is initialized to orientate the tracker in such a way that the sensor axes point in exactly the same direction as the axes of the operator's global coordinate frame. The sample frequency used is 100 Hz with a baudrate of 115 Kbps.

Two harmonic drive actuators are used to move the pan and tilt axis of the stereo head, based on feedback acquired from incremental position encoders. The chosen high precision encoders guarantee a specification 0.01 degree resolution and a maximum frequency response of 100 KHz. The DC servo motors have a maximum output speed of 110 rpm and maximum radial load 59 N, which is adequate for the two cameras load. Each servo is connected to a controller which sends low-level commands to the actuators for executing the trajectories received by the head tracker. A very precise calibration of the controllers was performed so that we could utilize the great degree precision of the position encoders. Position control strategy was chosen while position is the most important aspect of a high performance head tracking control. The Proportional Integral Derivative (PID) controller values were calibrated in discrete-time through the use of real-time processes running with fixed time steps. The use of a simple, and easy to tune control strategy across the pan and tilt axis helped to ensure the reliability and robustness of the whole system. The following equation represents the general PID controller (Astrom & Hagglund (1995)).

$$u = K_p e + K_i \int edt + K_d \frac{d(-PV)}{dt} \tag{1}$$

Position control requires an additional controller on top of the velocity controller since it sets the desired velocities in all driving phases, especially during the acceleration and deceleration phases. This control procedure has to take into account not only the current speed as a feedback value, but also the current position, since previous speed changes or inaccuracies may have had already an effect on the robot's position. The chosen experimental parameter tuning can be described by the following simple steps (Braunl (2008))

- Selection of the typical operating setting for the desired speed, set to zero integral and derivative parts, and then increase of K_p to maximum or until oscillation occurs.
- Division of K_p by two, when oscillation occurs.
- Slowly increase of K_d while increasing or decreasing the speed. For the smoothest response choose the selected value of K_d.
- Slowly increase K_i until oscillation starts. Then divide K_i by 2 or 3.
- In case the overall controller performance is satisfactorily under the typical system conditions, the tuning is successful.

3.2 Software Architecture

Key feature for the implementation of the real-time control is the used operating system. Since critical applications such as control need low response times, RT-Linux operating system is ideal for both MMU and MCU host computers. The distribution used is an Open Source project which provides an integrated execution environment for embedded real-time applications (Mantegazza et al. (2000)). It is based on components and incorporates the latest techniques for building embedded systems. The architecture is designed to develop hybrid systems with hard and soft real-time activities as shown in Fig. 4. The Linux kernel is treated as the lowest priority task under the RT kernel. In this case, we allocated the critical task of control at the RT-Linux level and the less critical tasks, such as inertial data filtering, at the Linux level. The real-time tasks need to communicate with user-space processes for things like file access, network communication or user interface. Thus, it provides FIFOs and shared memory implementations that provide communication with this user-space processes. The interface for both kinds of activities is a POSIX based interface. Software routines such as boot code, initialization positions, and input/output functions were developed using a combination of hand coded C or assembly language.

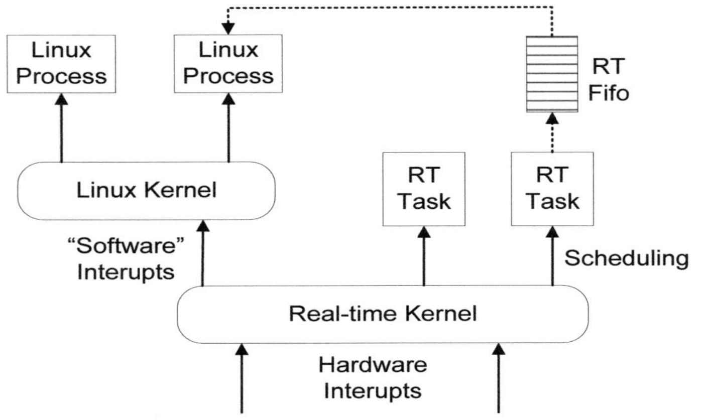

Fig. 4. The chosen operating system combines the use of two kernels, RT-Linux and Linux to provide support for critical tasks and soft real-time applications, respectively.

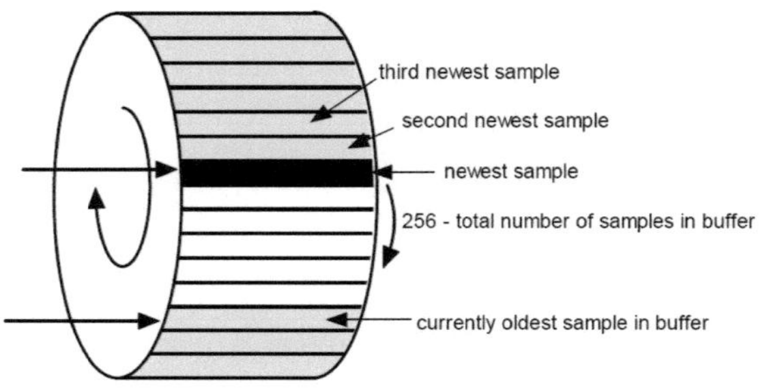

Fig. 5. The polling or event mechanism in the internal buffer of sensor.

The data received by the head tracker at the MCU, was firstly filtered by a Kalman filter. The software strategy dilemma of whether to use polling or events was considered in our implementation. Apart from the fact that the choice is mostly dependent on the user programming environment several other considerations were examined. When using the polling method, the user continuously or at a certain interval, queries the head tracker if new orientation data has been calculated. When queried, the sensor will immediately return the most recently calculated data, as shown in Fig. 5. The polling method is useful when the query function runs in a loop at a certain update rate and each time orientation data is needed, the user just needs the latest data and not necessarily every single sample. When using the events method, instead of continuously querying the sensor, the event notifies the user when new data has been calculated and is available for retrieval with the appropriate functions. For the presented system, the appropriate solution is the polling method since it ensures that the operator always get the latest available orientation data when he asks for it. The polling method allows that the other processes in our software to be asynchronous with the sampling rate of the head tracker itself, and we can synchronize the data with our processes. Furthermore, polling is slightly more straightforward to implement.

The errors in the force measurements introduced by our accelerometer and the errors in the measurement of angular change in the orientation with respect to the inertial space introduced by gyroscopes, were the two fundamental error sources which affected the error behavior of the operators head trajectory. Furthermore, all inertial measurements are corrupted by additive noise (Ovaska & Valiviita (1998)). The Kalman filter (Welch & Bishop (2001), Trucco & Verri (1998)) was used while is a form of optimal estimator, characterized by recursive evaluation using an estimated internal model of the dynamics of the system. The filtering was implemented on the MCU computer where the inertial sensor is attached, using the soft-real time kernel.

The control data received from the MMU, should be translated into motor commands for the equivalent axis. This operation is time critical since fast and accurate position commands to a remote robot is the only way to guarantee its safe operation. This strategy of considering the head tracker commands time critical and their implementation in the hard real-time kernel, allows the overall system to be flexible in a way that additional future motor commands and even more crucial, like the operation of a gripper, can be implemented easily while satisfying the hard real-time constraints.

The concurrency and parallelism was considered in the programming of the robotic system by using a multi-thread model. The motor run time models are not using the *wait.until.done*() function, while a change in the operator's field of view indicates that the previous movement should not be completed but a new motion position command should be addressed. The following runtime model was chosen for the motor class:

```
Thread 1 (control)
motor.change_position()
do other things

Thread 2 (monitor)
periodically wake up
read.new_sensor_position()
if (new_sensor_position !=...
...old_sensor_position)
old_sensor_position=motor.change_...
...position.(new_sensor_position)
```

Simultaneous and non-synchronized accesses to the same resources, such as servo motors, was not a set of problems for the implementation while the the pitch and yaw commands would move separately the tilt and pan axis, respectively. However, in case of a future additional operation, such as motor stabilization, the sharing of the same resources would be a great problem. Thus, the software programming infrastructure considered the shared resources and critical sections in order to guarantee the expandability and flexibility of the stereo vision system. The critical sections were easily implemented since the protected operations were limited. However, special attention must be paid since critical sections can disable system interrupts and can impact the responsiveness of the operating system.

4. Video Streaming System

4.1 Hardware Architecture

Each of the stereo head cameras on MMU is capable of outputting progressively images of 640×480 pixel resolution at maximum 30 frames per second. The digital cameras transmit the images over the fast USB 2.0 interface directly to the host's memory without the usage of frame grabbers. In order to determine the internal camera geometric and optical characteristics, camera calibration was necessary. A variety of methods have been reported in the bibliography. The method we used is described in (Bouget (2001)) using its available C Open Source code. The method is a non self-calibrating thus, we used a projected chessboard pattern to estimate the camera intrinsics and plane poses. Finally, the calibration results were used to rectify the images taken from cameras in order to have the best results in possible subsequent image processing algorithms. The video processing requires high computational burden and resources while it makes a full usage of certain instruction sets of a modern microprocessor. Thus, a high performance processor was chosen for the MMU computer. The operator in the MCU receives the stereo pair of images in the stereo HMD. The chosen HMD has the same input resolution like cameras 640×480 and a refresh rate of $70Hz$. Two separate 15 pin D-Sub (VGA) interfaces are used for the stereo image input to the HMD. Thus, the MCU computer is equipped with a double output high performance graphic card in order to display in different outputs each video stream.

4.2 Software Architecture

Vision systems of mobile robots must unify the requirements and demands of both computer vision and image processing disciplines and robotic and embedded system disciplines. While the state of the art in computer vision algorithms is advanced, many computer vision processes are computationally expensive and thus inappropriately for real-time applications. Therefore, the resource demands of computer vision applications are in conflict with the requirements posed by robotics and embedded systems. For this system, a compression scheme must be implemented in order to transmit the stereo image stream. The high input data rate from the cameras of $2(stereo) \times 640 \times 480(resolution) \times 3(color) \times 8(bit\,per\,pixel) \times 25(fps) \cong$ 351 Mbps requires a compression algorithm with high compression ratio, low computational complexity and good output quality. Furthermore, the compressed video should be packetized and streamed over the communication network.

The architecture chosen aims to make the MMU computer a video server which will perform the following primary tasks:

- Capture video from both cameras
- Compress video using a codec
- Packetize the compressed video and attach time stamps within the packets
- Stream the packets over the communication network

For all the previous tasks, apart from capturing, the FFmpeg video open source libraries (FFmpeg project (2008)) were selected. The video server allows multicast transmission while it sends each video stream to a fixed-destination multicast address. The services dealing with each stream, like the video player in the MCU, only have to listen to the appropriate multicast address, so several services can receive the same video stream without increasing bandwidth consumption. The compression was done using MPEG-4 codec, and the transmission of the video streams using the MPEG Transport Stream (Gringeri et al. (1998)). MPEG-TS provides many features found in data link layers, such as packet identification, synchronization, timing (clock references and timestamps), multiplexing and sequencing information. In the architecture chosen, each processing tree is executed within its own thread and is processed in parallel with other source nodes, like the control loop. This framework ensures appropriate synchronization between the image streams. With this framework, the developers do not need to worry about locking issues and synchronization primitives. The UDP communication protocol was used between the two computers while it uses a higher level abstraction, it is built on top of sockets, and meets the requirements for low latency (Traylor et al. (2005)).

Fig. 6. The remotely operated robot prototype.

For video capturing the video for Linux (Video 4 Linux project (2008)) driver was chosen, which is an open source application programming interface for video capture and output drivers. The available streaming parameters were used to optimize the video capture process as well as the I/O. Our pre-selected video options determine a default number of frames per second in both digital cameras. If less than this number of frames is to be captured or output, applications can request frame skipping or duplicating on the driver side. This is especially useful when using the priority handling of the wireless network topology, where cases of video frames not augmented by timestamps or sequence counters are necessary for bandwidth saving. In order to exchange images between drivers and applications, it was necessary to have standard image data formats which both sides will interpret the same way. The used interface included several such formats but it was not limited only to these formats since driver-specific formats were possible since in the presented stereo vision system, some applications depended on codecs to convert images to one of the standard formats when needed.
The I/O streaming was designed in a way that only pointers to buffers were exchanged between application and driver, ensuring that the data itself was not copied. The capturing application enqueued a number of empty buffers before starting capturing and entering the read loop. The application waited until a filled buffer could be dequeued, and re-enqueued the buffer when the data was no longer needed. Output applications filled and enqueued

the buffers, and when enough buffers were stacked up output was started. In the write loop, when the application run out of free buffers it waited until an empty buffer could be dequeued and reused.

One of the highlights of the presented system is the multiple concepts for real-time image processing. Each image processing tree is executed with its own thread priority and scheduler choice, which is directly mapped to the operating system process scheduler. This was necessary in order to minimize jitter and ensure correct priorization, especially under heavy load situations. Some of the performed image processing tasks were disparity estimation (Georgoulas et al. (2008)), object tracking (Metta et al. (2004)), image stabilization (Amanatiadis et al. (2007)) and image zooming (Amanatiadis & Andreadis (2008)). For all these image processing cases, a careful selection of programming platform should be made. Thus, the open source computer vision library, OpenCV, was chosen for our image processing algorithms (Bradski & Kaehler (2008)). OpenCV was designed for computational efficiency and with a strong focus on real-time applications. It is written in optimized C and can take advantage of multicore processors. The basic components in the library were complete enough to enable the creation of our solutions.

In the MCU computer, an open source video player VLC was chosen for the playback service of the video streams VideoLAN project (2008). VLC is an open source cross-platform media player which supports a large number of multimedia formats and it is based on the FFmpeg libraries. The same FFmpeg libraries are now decoding and synchronize the received UDP packets. Two different instances of the player are functioning in different network ports. Each stream from the video server is transmitted to the same network address, the MCU network address, but in different ports. Thus, each player receives the right stream and with the help of the MCU on board graphic card capabilities, each stream is directed to one of the two available VGA inputs of the HMD.

The above chosen architecture offers a great flexibility and expandability in many different aspects. In the MMU, additional video camera devices can be easily added and be attached to the video server. Image processing algorithms and effects can be implemented using the open source video libraries like filtering, scaling and overlaying. Furthermore, in the MCU, additional video clients can be added easily and controlled separately.

5. System Performance

Laboratory testing and extensive open field tests, as shown in Fig. 6, have been carried out in order to evaluate the overall system performance. During calibration of the PIDs the chosen gains of (1) were $K_p = 58.6$, $K_i = 2000$ and $K_d = 340.2$. The aim of the controlling architecture was to guarantee the fine response and accurate axis movement. Figure 7 shows the response of the position controller in internal units (IU). One degree equals to 640 IU of the encoder. As we can see, the position controller has a very good response and follows the target position. From the position error plot we can determine that the maximum error is 17 IU which equals to 0.026 degrees.

To confirm the validity of the vision system architecture scheme, of selecting RT-Linux kernel operating system for the control commands, interrupt latency was measured on a PC which has an Athlon 1.2GHz processor. In order to assess the effect of the operating system latency, we ran an I/O stress test as a competing background load while running the control commands. With this background running, a thread fetched the CPU clock-count and issued a control command, which caused the interrupt; triggered by the interrupt, an interrupt handler (another thread) got the CPU clock-count again and cleared the interrupt. Iterating the

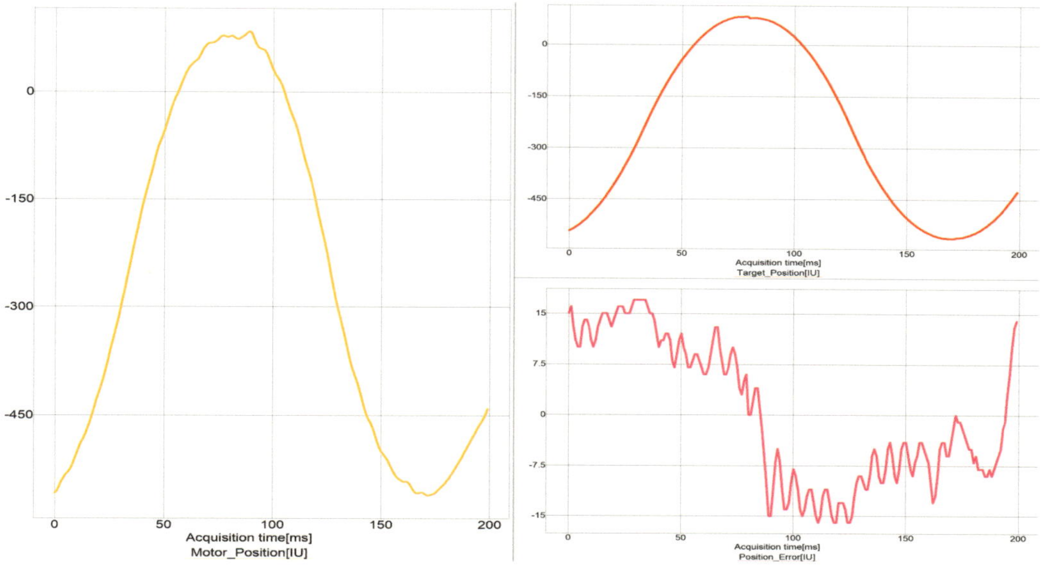

Fig. 7. A plot of position controller performance. Left: The motor position, Up Right: The target motor position, Down Right: The position error.

above steps, the latency, the difference of the two clock-count values, was measured. On standard Linux kernel, the maximum latency was more than 400 msec, with a large variance in the measures. In the stereo vision system implementation in RT-Linux kernel the latency was significantly lower with maximum latency less than 30 msec and very low variation.

The third set of results show the inter-frame times, the difference between the display times of a video frame and the previous frame. The expected inter-frame time is the process period $1/f$ where f is the video frame rate. In our experiments, we used the VLC player for the playback in the MMU host computer. We chose to make the measurements on the MMU and not on the MCU computer in order to calculate only the operating system latency avoiding overheads from communication protocol latencies and priorities. The selected video frame rate was 30 frames per second. Thus, the expected inter-frame time was 33.3 msec. Figure 8(a) shows the inter-frame times obtained using only the standard Linux kernel for both control and video process. The measurements were taken with heavy control commands running in the background. The inter-frame time due to the control process load introduces additional variation in the inter-frame times and increases these times to more than 40ms. In contrast, Figure 8(b) shows the inter-frame times obtained using the RT-Linux kernel with high resolution timers for the control process and the standard Linux kernel for the video process. The measurements were taken with the same heavy control commands running in the background. As we can see, the inter-frame times are clustered more around the correct value of 33.3 msec and their variation is lower.

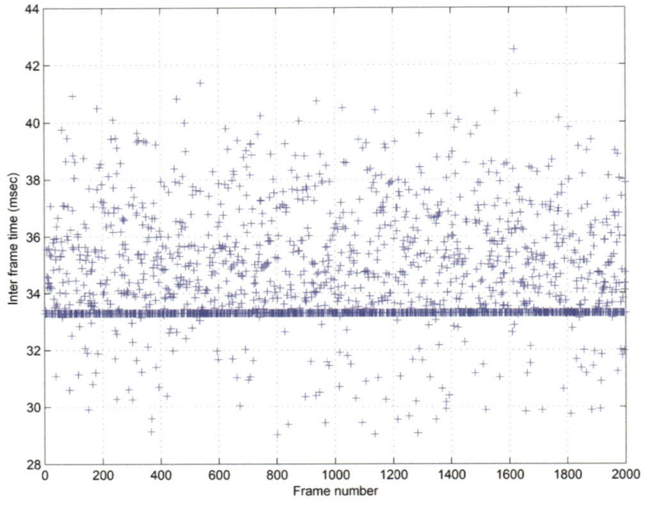

(a)

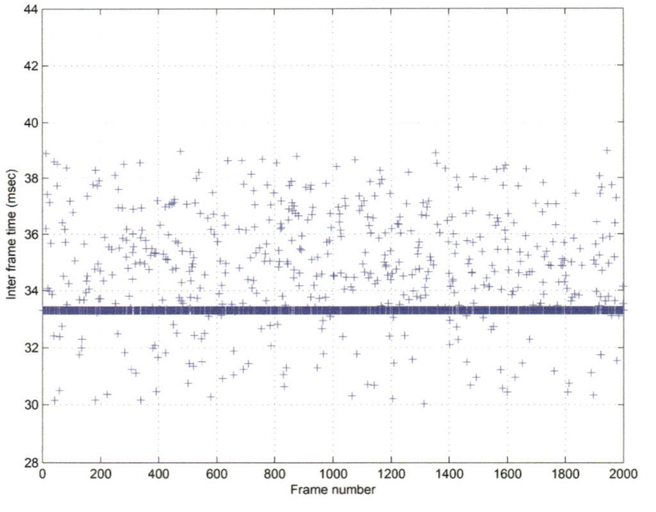

(b)

Fig. 8. Inter-frame time measurements: (a) Both control and video process running in standard Linux kernel; (b) Control process running in RT-Linux kernel and video process in standard Linux kernel.

6. Conclusion

This chapter described a robust prototype stereo vision paradigm for real-time applications, based on open source libraries. The system was designed and implemented to serve as a binocular head for remotely operated robots. The two main implemented processes were the remote control of the head via a head tracker and the stereo video streaming to the mobile control unit. The key features of the design of the stereo vision system include:

- A complete implementation with the use of open source libraries based on two RT-Linux operating systems
- A hard real-time implementation for the control commands
- A low latency implementation for the video streaming transmission
- A flexible and easily expandable control and video streaming architecture for future improvements and additions

All the aforementioned features make the presented implementation appropriate for sophisticated remotely operated robots.

7. References

Amanatiadis, A. & Andreadis, I. (2008). An integrated architecture for adaptive image stabilization in zooming operation, *IEEE Transactions on Consumer Electronics* **54**(2): 600–608.

Amanatiadis, A., Andreadis, I., Gasteratos, A. & Kyriakoulis, N. (2007). A rotational and translational image stabilization system for remotely operated robots, *Proc. of the IEEE Int. Workshop on Imaging Systems and Techniques*, pp. 1–5.

Astrom, K. & Hagglund, T. (1995). *PID controllers: Theory, Design and Tuning*, Instrument Society of America, Research Triangle Park.

Bluethmann, W., Ambrose, R., Diftler, M., Askew, S., Huber, E., Goza, M., Rehnmark, F., Lovchik, C. & Magruder, D. (2003). Robonaut: A robot designed to work with humans in space, *Autonomous Robots* **14**(2): 179–197.

Bouget, J. (2001). Camera calibration toolbox for Matlab, *California Institute of Technology, http//www.vision.caltech.edu* .

Bradski, G. & Kaehler, A. (2008). *Learning OpenCV: Computer vision with the OpenCV library*, O'Reilly Media, Inc.

Braunl, T. (2008). *Embedded robotics: mobile robot design and applications with embedded systems*, Springer-Verlag New York Inc.

Davids, A. (2002). Urban search and rescue robots: from tragedy to technology, *IEEE Intell. Syst.* **17**(2): 81–83.

Desouza, G. & Kak, A. (2002). Vision for mobile robot navigation: a survey, *IEEE Trans. Pattern Anal. Mach. Intell.* **24**(2): 237–267.

FFmpeg project (2008). *http//ffmpeg.sourceforge.net* .

Fong, T. & Thorpe, C. (2001). Vehicle teleoperation interfaces, *Autonomous Robots* **11**(1): 9–18.

Georgoulas, C., Kotoulas, L., Sirakoulis, G., Andreadis, I. & Gasteratos, A. (2008). Real-time disparity map computation module, *Microprocessors and Microsystems* **32**(3): 159–170.

Gringeri, S., Khasnabish, B., Lewis, A., Shuaib, K., Egorov, R. & Basch, B. (1998). Transmission of MPEG-2 video streams over ATM, *IEEE Multimedia* **5**(1): 58–71.

Kofman, J., Wu, X., Luu, T. & Verma, S. (2005). Teleoperation of a robot manipulator using a vision-based human-robot interface, *IEEE Trans. Ind. Electron.* **52**(5): 1206–1219.

Mantegazza, P., Dozio, E. & Papacharalambous, S. (2000). RTAI: Real time application interface, *Linux Journal* **2000**(72es).

Marin, R., Sanz, P., Nebot, P. & Wirz, R. (2005). A multimodal interface to control a robot arm via the web: a case study on remote programming, *IEEE Trans. Ind. Electron.* **52**(6): 1506–1520.

Metta, G., Gasteratos, A. & Sandini, G. (2004). Learning to track colored objects with log-polar vision, *Mechatronics* **14**(9): 989–1006.

Murphy, R. (2004). Human-robot interaction in rescue robotics, *IEEE Trans. Syst., Man, Cybern., Part C,* **34**(2): 138–153.

Ovaska, S. & Valiviita, S. (1998). Angular acceleration measurement: A review, *IEEE Trans. Instrum. Meas.* **47**(5): 1211–1217.

Roetenberg, D., Luinge, H., Baten, C. & Veltink, P. (2005). Compensation of magnetic disturbances improves inertial and magnetic sensing of human body segment orientation, *IEEE Transactions on neural systems and rehabilitation engineering* **13**(3): 395–405.

Tachi, S., Komoriya, K., Sawada, K., Nishiyama, T., Itoko, T., Kobayashi, M. & Inoue, K. (2003). Telexistence cockpit for humanoid robot control, *Advanced Robotics* **17**(3): 199–217.

Traylor, R., Wilhelm, D., Adelstein, B. & Tan, H. (2005). Design considerations for stand-alone haptic interfaces communicating via UDP protocol, *Proceedings of the 2005 World Haptics Conference*, pp. 563–564.

Trucco, E. & Verri, A. (1998). *Introductory Techniques for 3-D Computer Vision*, Prentice Hall PTR Upper Saddle River, NJ, USA.

Video 4 Linux project (2008). *http://linuxtv.org/* .

VideoLAN project (2008). *http://www.videolan.org/* .

Welch, G. & Bishop, G. (2001). An introduction to the Kalman filter, *ACM SIGGRAPH 2001 Course Notes* .

Willemsen, P., Colton, M., Creem-Regehr, S. & Thompson, W. (2004). The effects of head-mounted display mechanics on distance judgments in virtual environments, *Proc. of the 1st Symposium on Applied perception in graphics and visualization*, pp. 35–38.

5

Virtual Ubiquitous Robotic Space and Its Network-based Services

Kyeong-Won Jeon*,**, Yong-Moo Kwon* and Hanseok Ko**
Korea Institute of Science & Technology, Korea University***
Korea

1. Introduction

A ubiquitous robotic space (URS) refers to a special kind of environment in which robots gain enhanced perception, recognition, decision, and execution capabilities through distributed sensing and computing, thus responding intelligently to the needs of humans and current context of the space. The URS also aims to build a smart environment by developing a generic framework in which a plurality of technologies including robotics, network and communications can be integrated synergistically. The URS comprises three spaces: physical, semantic, and virtual space (Wonpil Yu, Jae-Yeong Lee, Young-Guk Ha, Minsu Jang, Joo-Chan Sohn, Yong-Moo Kwon, and Hyo-Sung Ahn Oct. 2009).

This chapter introduces the concept of virtual URS and its network-based services. The primary role of the virtual URS is to provide users with a 2D or 3D virtual model of the physical space, thereby enabling the user to investigate and interact with the physical space in an intuitive way. The motivation of virtual URS is to create new services by combining robot and VR (virtual reality) technologies together.

The chapter is composed of three parts: what is the virtual URS, how to model virtual URS and its network-based services.

The first part describes the concept of virtual URS. The virtual URS is a virtual space for intuitive human-robotic space interface, which provides geometry and texture information of the corresponding physical space. The virtual URS is the intermediate between the real robot space and human. It can represent the status of physical URS, e.g., robot position and real environment sensor position/status based on 2D/3D indoor model.

The second part describes modeling of indoor space and environment sensor for the virtual URS.
There were several researches for indoor space geometry modeling (Liu, R. Emery, D.Chakrabarti, W. Burgard and S. Thrun 2001), (Hahnel, W. Burgard, and S. Thrun (July, 2003), (Peter Biber, Henrik Andreasson, Tom Duckett, and Andreas Schilling, et al. 2004).

Here, we will introduce our simple and easy to use indoor modeling method using 2D LRF (Laser Range Finder) and camera. For supporting web service, VRML and SVG techniques are applied. In case of environment sensor modeling, XML technology is applied while coordination with the web service technologies. As an example, several sensors (temperature, light, RFID etc) of indoor are modeled and managed in the web server.

The third part describes network-based virtual URS applications: indoor surveillance and sensor-based environment monitoring. These services can be provided through internet web browser and mobile phone.

In case of indoor surveillance, the human-robot interaction service using the virtual URS is described. Especially, the mobile phone based 3D indoor model browsing and tele-operation of robot are described.

In case of sensor-responsive environment monitoring, the concept of sensor-responsive virtual URS is described. In more detail, several implementation issues on the sensor data acquisition, communication, 3D web and visualization techniques are described. A demonstration example of sensor-responsive virtual URS is introduced.

2. Virtual Ubiquitous Robotic Space

2.1 Concpet of Virtual URS

The virtual URS is a virtual space for intuitive human-URS (or robot) interface, which provides geometry and texture information of the corresponding physical space. Fig. 1 shows the concept of virtual URS which is an intuitive interface between human and physical URS. In physical URS, there may be robot and ubiquitous sensor network (USN) which are real things in our close indoor environment. For example, the robot can perform the security duty and sensor network information is updated to be used as a decision ground of whole devices' operation. The virtual URS is the intermediate between the real robot space and human. It can represent the status of physical URS, e.g., robot position and sensor position/status based on 2D/3D indoor model.

2.2 Concept of Responsive Virtual URS

The virtual URS can be responded according to the sensor status. We construct sensor network in virtual URS and define the space as a responsive virtual URS. In other words, responsive virtual URS is generated by modeling of indoor space and sensors. So sensor status is reflected to space. As a simple example, the light rendering in virtual URS can be changed according to the light sensor information in physical space. This is a concept of responsive virtual URS which provides similar environment model to the corresponding physical URS status. In other words, when event happens in physical URS, the virtual URS responds. Fig. 2 shows that the responsive virtual URS is based on indoor modeling and sensor modeling.

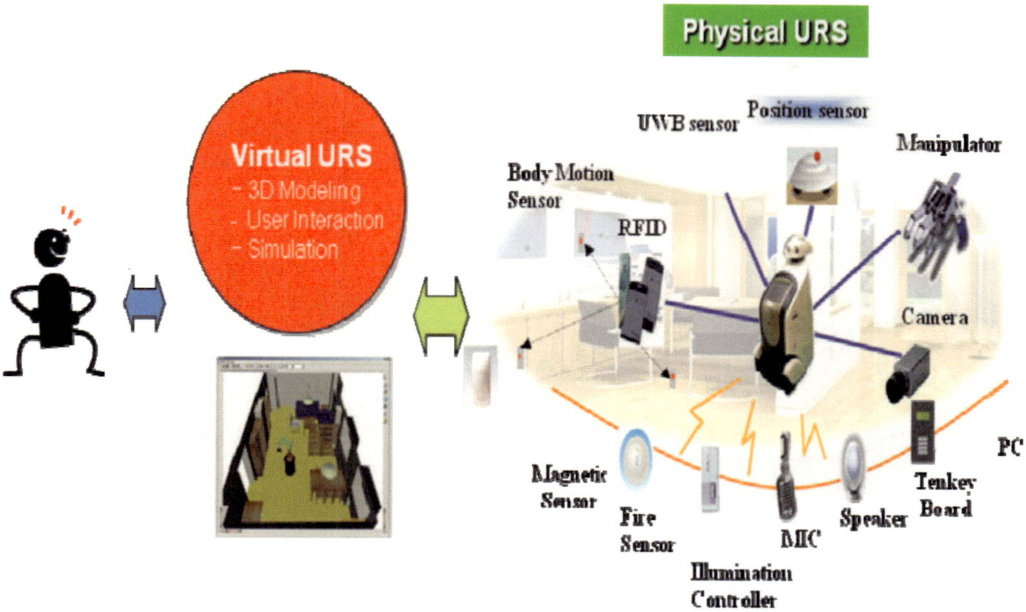

Fig. 1. Concept of physical and virtual URS

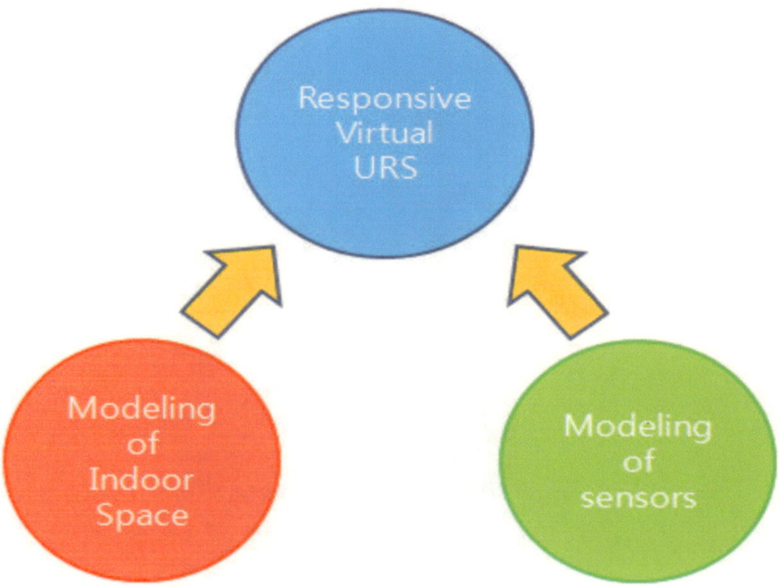

Fig. 2. The concept of responsive virtual URS

3. Modeling Issues

3.1 Modeling of Indoor Space

This section gives an overview of our method to build a 3D model of an indoor environment. Fig. 3 shows our approach for indoor modeling. As shown in Fig. 3, there are three steps, localization of data acuqisition device, acquisition of geometry data, texture image capturing and mapping to geometry model.

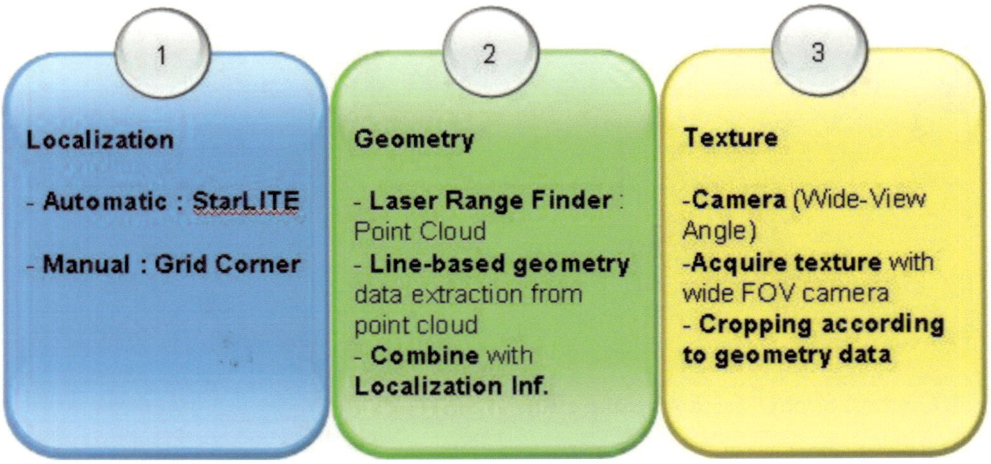

Fig. 3. Indoor modeling process

Fig. 4. Indoor 3D modeling platform

The localization information is used for building the overall indoor model. In our research, we use two approaches. One is using IR landmark-based localization device, named as starLITE (Heeseoung Chae, Jaeyeong Lee and Wonpil Yu 2005), the other is using dimension of floor square tile (DFST) manually. The starLITE approach can be used automatic localization. The DFST approach is applied when starLITE is not installed. The DFST method can be used easily in the environment that has reference dimension without the additional cost for the localization device, although it takes times due to the manual localization.

In case of 2D model & 3D model, the geometry data is acquired with 2D laser scanner. In case of 2D laser scanner, we used two kinds of laser scanners, i.e., SICK LMS 200 and Hokuyo URG laser range finder.

Fig. 4 shows our indoor modeling platform using two Hokuyo URG laser range finder (LRF) and one IEEE-1394 camera. One scans indoor environment horizontally and another scans indoor environment vertically. From each LRF, we can generate 2D geometry data by gathering and merging point clouds data. Then, we can get 3D geometry information by merging two LRF 2D geometry data. For texture, the aligned camera is used to capture texture images. Image warping, stitching and cropping operations are applied to texture images.

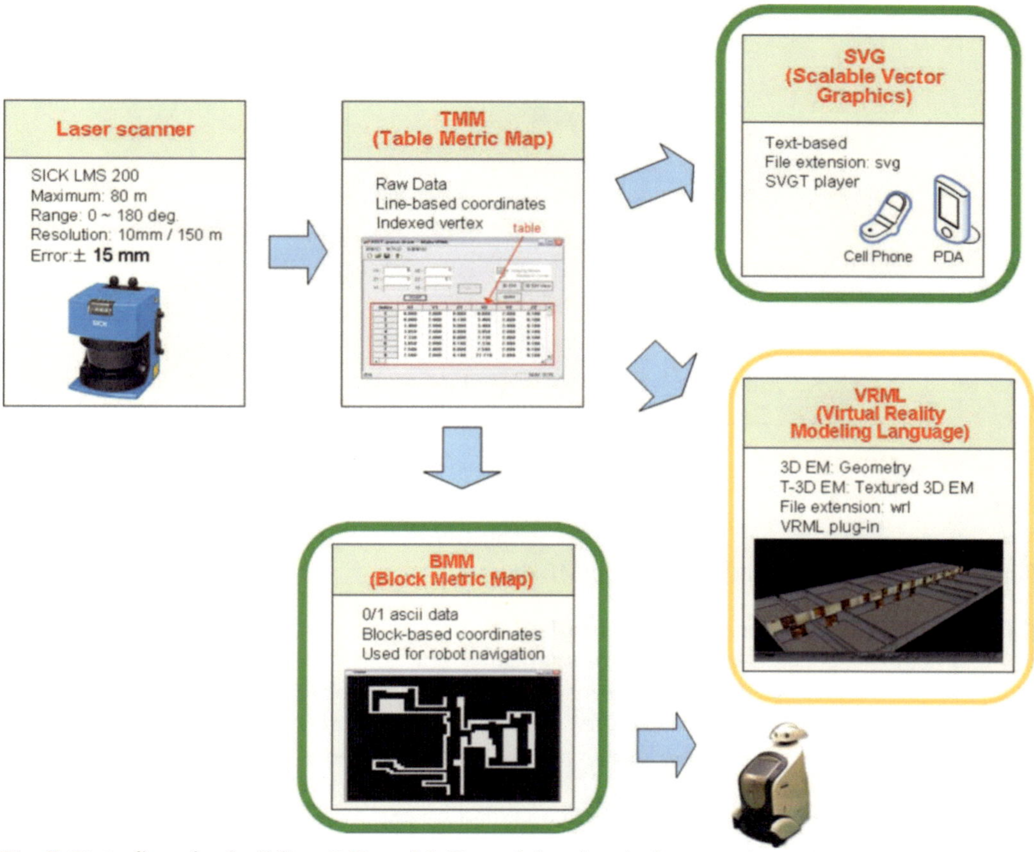

Fig. 5. Data flow for building 2-D and 3-D models of an indoor environment

Fig. 5 shows data flow for building 2-D and 3-D models of an indoor environment. Based on these geometry data acquired from laser scanner, TMM (Table Metric Map), GMM (Graphic Metric Map) and 3D EM (Environment Map) are generated automatically. It is also possible to apply same procedure if there is a metric flat drawing. In this case, the data acquisition with laser scanner is not needed. For the web service, the indoor model is generated based on SVG (2D) and VRML (3D). BMM (Block Metric Map) is also generated for the robot navigation map.

3.2 Modeling of Environment Sensor

(1) Sensor modeling based on XML

We implement XML based environment sensor modeling. Especially we design sensor XML GUI and develop XML sensor file generator according to the input data from GUI. As an example, the sensors we use are light, fire and gas sensor. We design the model data of sensors with sensor id, type, location and status.

Using our sensor XML GUI shown in Fig. 6, user can create the XML file for each sensor. In other words, user can input sensor id, type, location and status using GUI and then the corresponding XML sensor file is generated. For same sensor type, We can describe many sensors while using different sensor ids.

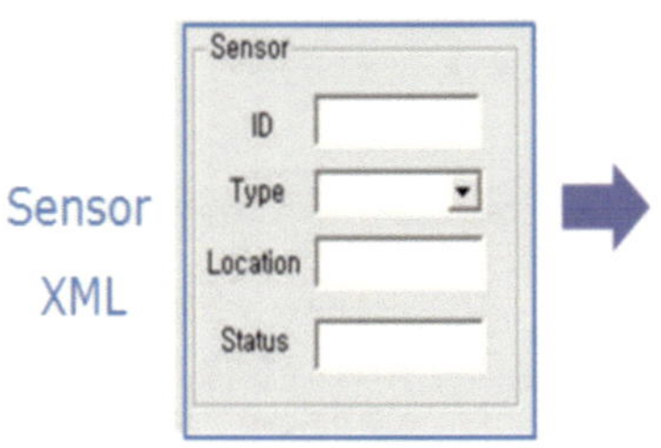


```
<?xml version="1.0" encoding="euc-kr" ?>
- <sensors>
  - <sensor>
      <id>FS1</id>
      <type>fire</type>
      <location>27,4,2.6</location>
      <status>fire on</status>
    </sensor>
  - <sensor>
      <id>LS1</id>
      <type>light</type>
      <location>27.4,4,2.6</location>
      <status>40</status>
    </sensor>
```


Fig. 6. Sensor XML GUI and generated XML data

(2) Automatic acquisition of sensor installation

We can acquire sensor information automatically. The sensor network is based on zigbee network. Sensor base station detects sensor which is now working and sensor information (sensor id, sensor type, sensor status etc.) based on data logging. So user is able to confirm the sensors that are working and know their id, name and value. All information are saved and managed by XML.

Fig. 7 shows automatic detection of newly installed sensor and addition of new sensor XML data into the previous XML sensor file.

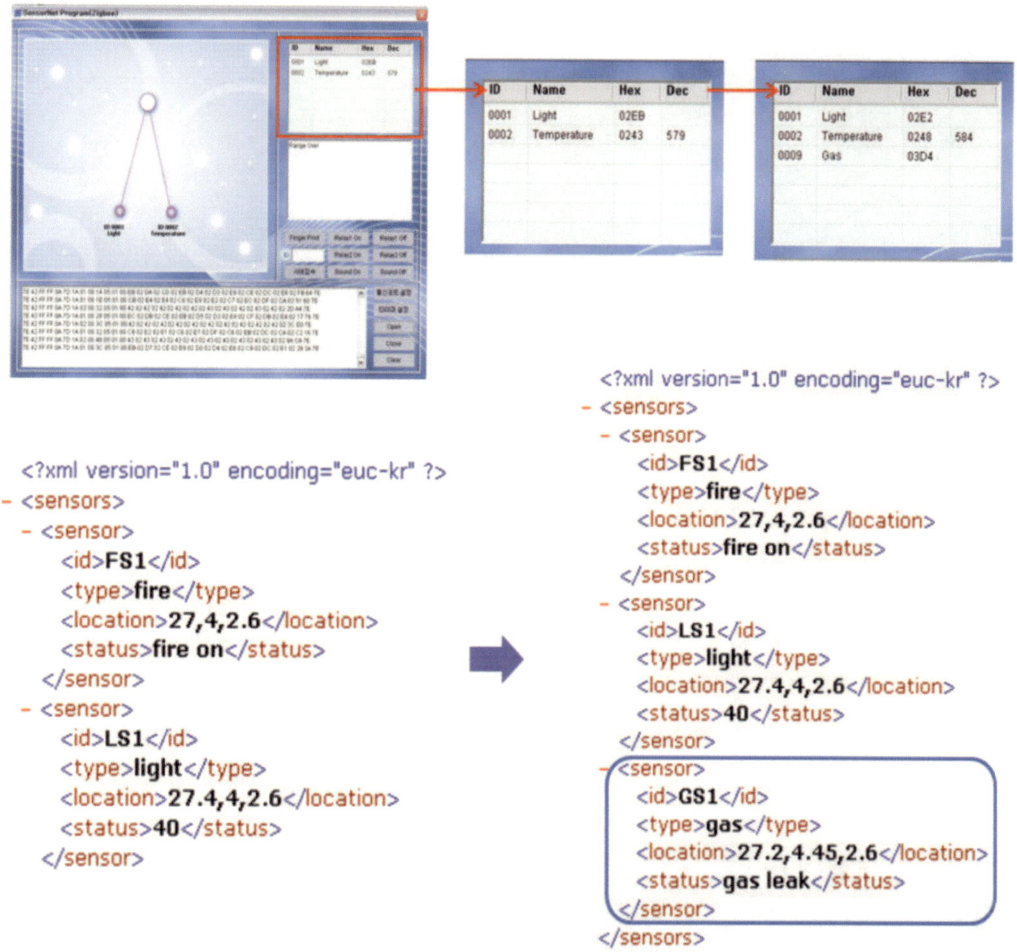

Fig. 7. Detection of sensor module and automatic addition of sensor model data in XML file

(3) Registration of sensor location

It should be noted that when we make a model for each sensor installed, the sensor id, type and status can be inserted automatically in XML sensor file. However, the location information of sensor is not easy to insert automatically, because we need to measure 3D location of sensor.

In this paper, we implement the virtual URS based input of sensor location information into XML file. Here, we assume that sensor is installed at ceiling and the height information is measured once and we know it. User can point out the corresponding sensor location roughly using the virtual URS (2D map or bird eye view 3D model). Then, the (x,y) data of sensor location can be extracted automatically and merged with the known height data for 3D location data of sensor. This 3D location data is saved to XML file automatically and virtual sensor is generated in the virtual URS according to XML file. Fig. 8 shows the basic concept of the virtual URS based insertion of sensor location data into XML file.

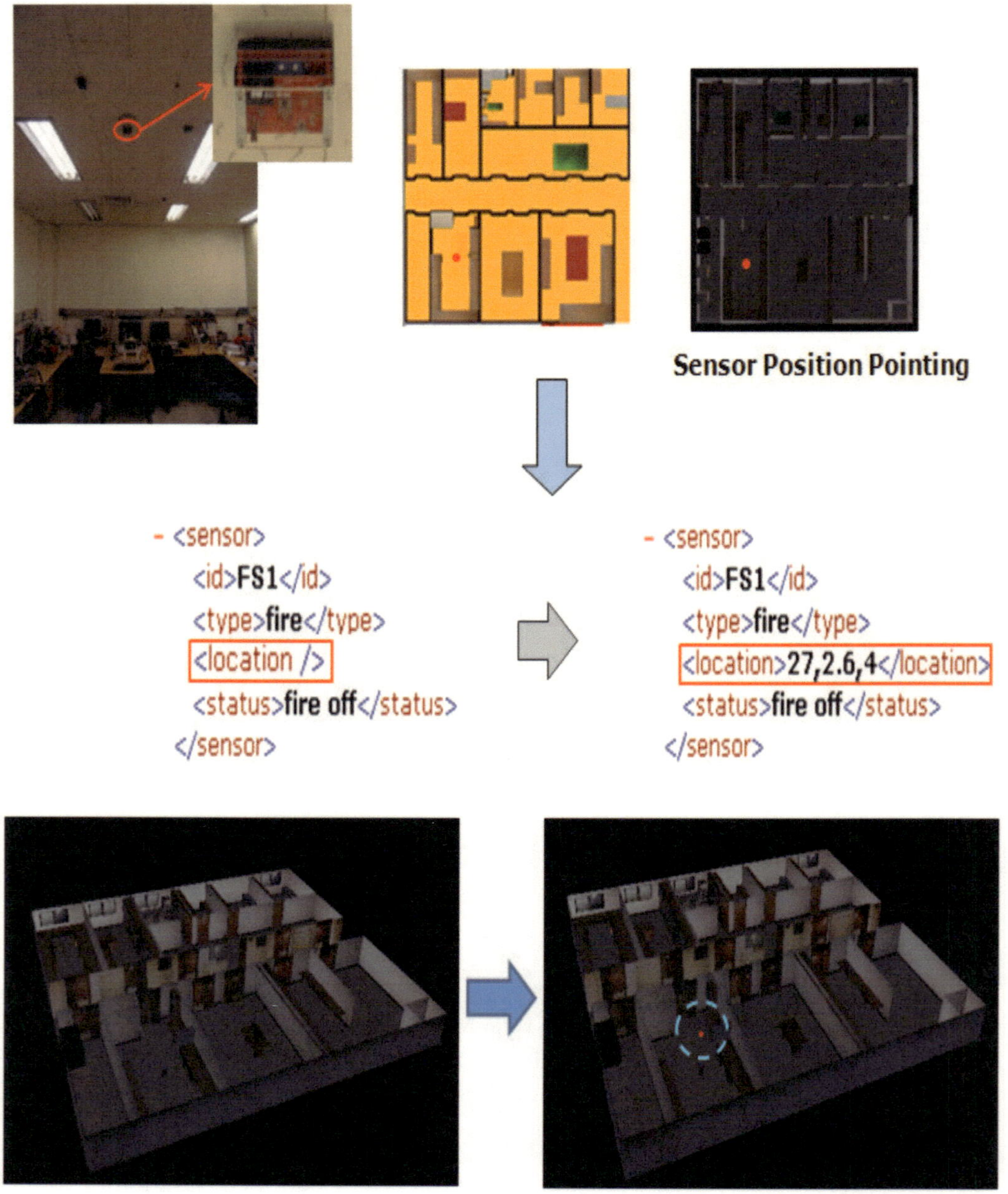

Fig. 8. Virtual URS-based insertion of sensor location to XML file

4. Virtual URS Services

4.1 Network-based Human-Robot Interaction

User can interact with robot through web server as shown in Fig. 9. For example, user can designate the destination point of robot and also receive the current robot position information through network. For interaction with robot, user can use several kinds of terminals like PC, PDA or mobile phone. Here, web server provides services of human-robot interaction.

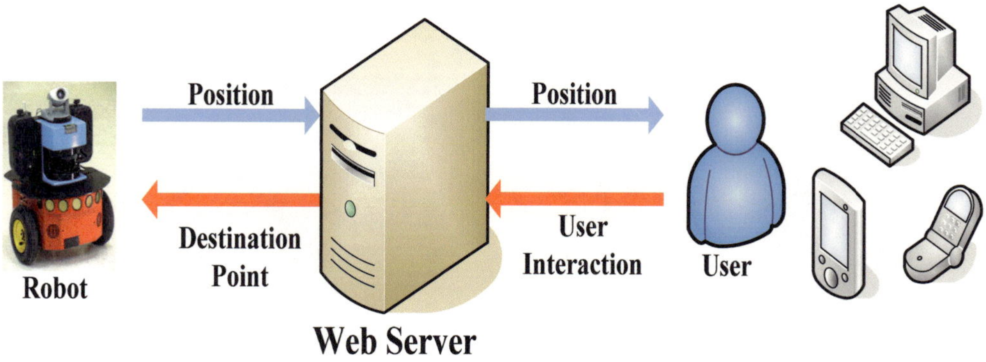

Fig. 9. Basic concept of network-based human-robot interaction

Here, we introduce two kinds of network-based human-robot interaction (HRI) services, i.e., HRI through web-browser and mobile phone. Fig. 10 shows overview of network-based human-robot interaction service system.

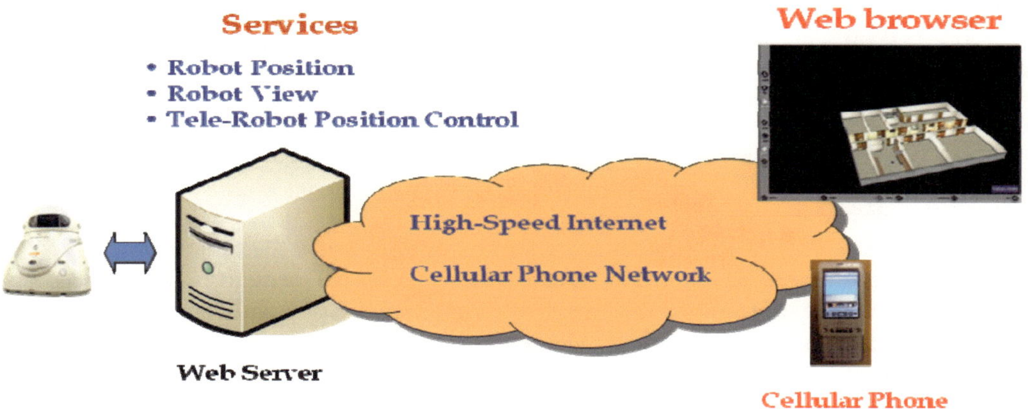

Fig. 10. Overview of network-based human-robot interaction service system

(1) Web browser -based interactive service

Basically, the virtual URS provides 2D/3D Model of URS, which supports indoor space model, object model and status update of robot or some object in space according to event occurred in physical URS. The virtual URS also provides the function such as display of robot location of physical URS, designating point and robot path planning. Through bridging between the virtual URS and the physical URS, user is able to command robot to move. Moreover, the user can pick many destinations to decide the robot path so that the robot will move according to the designated path. The functions are possible in remote environment through web.

Fig. 11 shows web browser-based interactive tele-presence through network, for example, tele-presence between KIST in Seoul, Korea and ETRI in Daejon, Korea. As shown in Fig. 11, our system can provide telepresence with the virtual URS including 3D indoor model, robot, sensor and map information. Moreover, user can interact with remote site robot through the virtual URS.

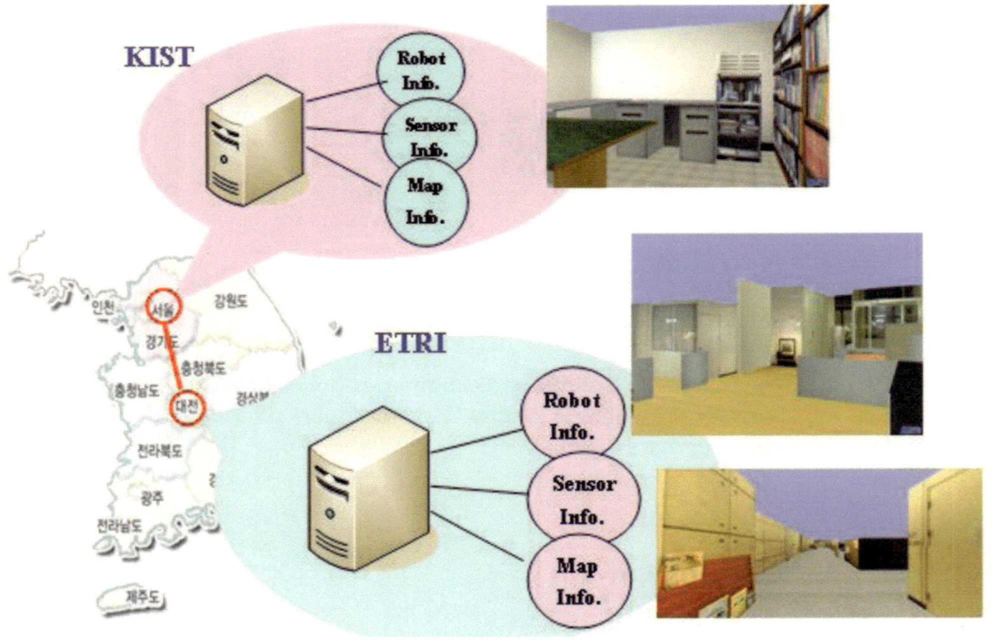

Fig. 11. Web browser-based interactive tele-presence

(2) Mobile phone-based interactive service

If user designates robot position in the virtual URS through web, the remote physical robot moves to the designated location. This function is also possible through mobile phone. It is possible that user can see the robot position, robot view in the physical URS through the 3D virtual URS while using mobile phone. Moreover user can control the robot in physical URS on mobile phone.

The 3D robot view service is impossible on general mobile phone without 3D engine. So we design a service platform for 3D mobile phone service (Kyeong-Won Jeon, Yong-Moo Kwon, Hanseok Ko, 2007). The service platform for the 3D virtual URS service on mobile phone is composed of 3D model server, 3D view image generation, mobile server and mobile phone.

- The 3D model server manages 3D model (VRML). Several 3D models exist in 3D model server.
- The 3D view image generation part is composed of 3D model browser and 3D model to 2D image converting program. 3D model browser is to render 3D view in 3D model. So user can see 3D view through the 3D model browser. Then, the rendered image is converted to 2D image (jpg).

- The mobile server manages the saved 2D image and sends it to mobile phone by TCP/IP communication. Another role of mobile server is to transfer interaction information from mobile phone to 3D model browser for interaction service between mobile phone and 3D model browser

Fig. 12 shows the architecture of mobile phone-based interactive service. Fig. 13 shows mobile phone based interaction to the virtual URS. Fig. 14 shows 3D view image on the mobile phone.

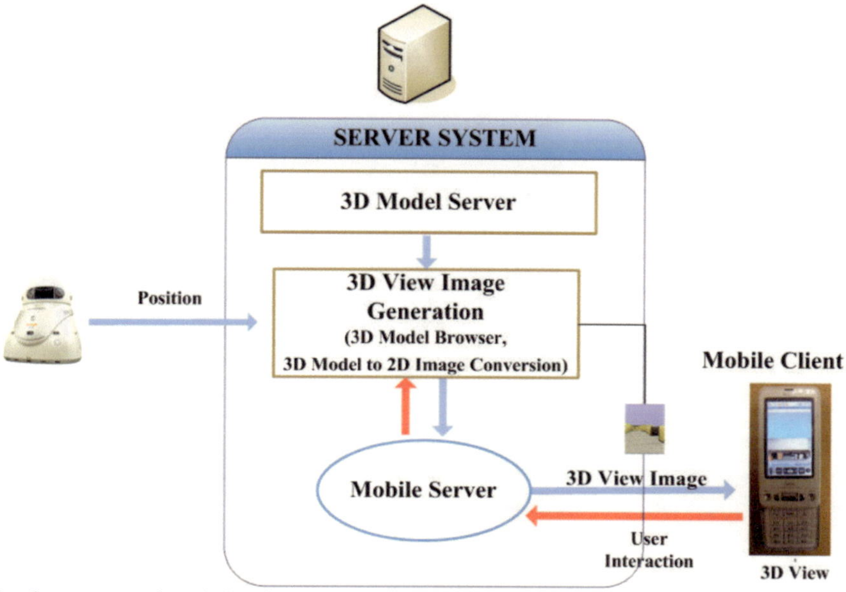

Fig. 12. Architecture of mobile phone-based interactive service

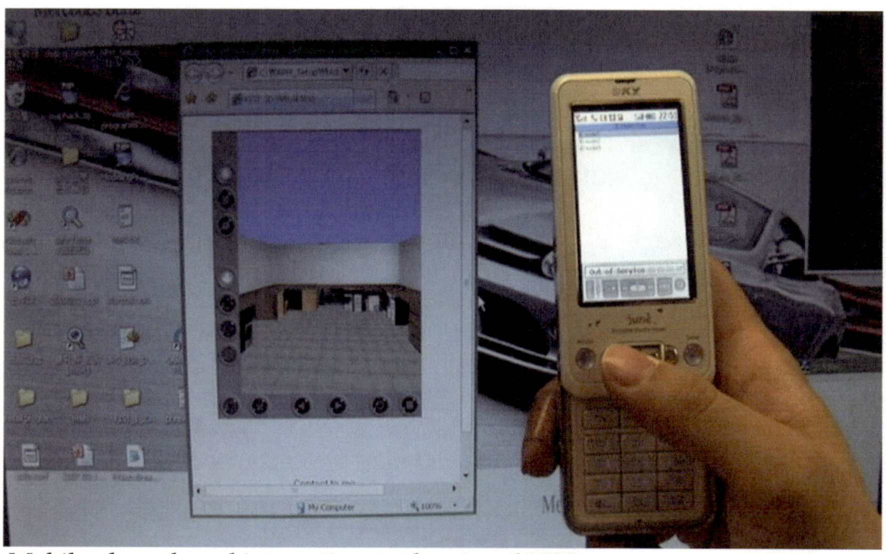

Fig. 13. Mobile phone based interaction to the virtual URS

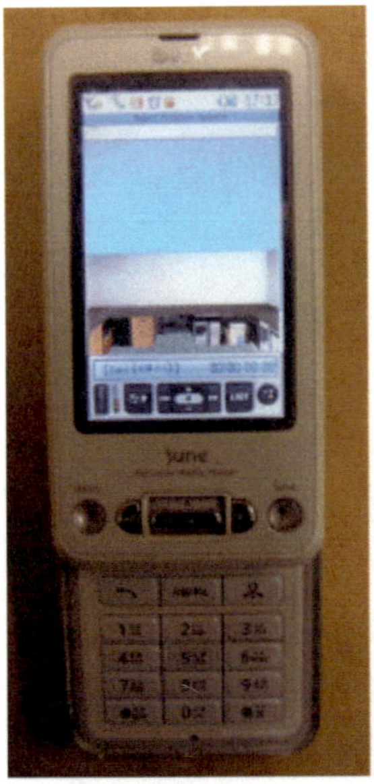

Fig. 14. 3D view image on the mobile phone

4.2 Sensor-Responsive Virtual URS

We provide sensor-responsive virtual URS service by bridging between the physical URS and the virtual URS. When an event happens in physical space, the sensor catches the event. Then the sensor id, sensor status information are delivered to the web server through the wireless network (for example, zigbee network). Upon receiving sensor status change information, the XML data is also updated automatically. In case of the robot position, it is continuously detected by sensor and then the XML robot data (robot position information) is updated. The XML robot data is reflected to robot in the virtual URS. Here, the XML file acts like a virtual sensor in the virtual URS. Then, the virtual URS also responds according to the virtual sensor status.

For example, if the status of fire sensor is activated, this information is transferred to the virtual URS and then the fire status in XML data is changed. Fig. 15 shows an automatic robot sensor status update in XML file.

The merit of VR technology is that user can experience virtually without experiencing actually. Because the virtual URS provides visual service, user can feel realistically by virtual experience. That is, visualization of the situation of physical URS is the role of virtual URS. User can confirm the status and position of robot and the situation of environment. Moreover, when event happens, robot view service is possible according to robot movement. Fig. 16 shows XML-based bridging between the physical URS and the virtual URS. Fig. 17 shows visualization service of senor in the virtual URS. Fig. 18 and Fig. 19 show a responsive virtual URSs accoring to the fire and light sensors, repectively.

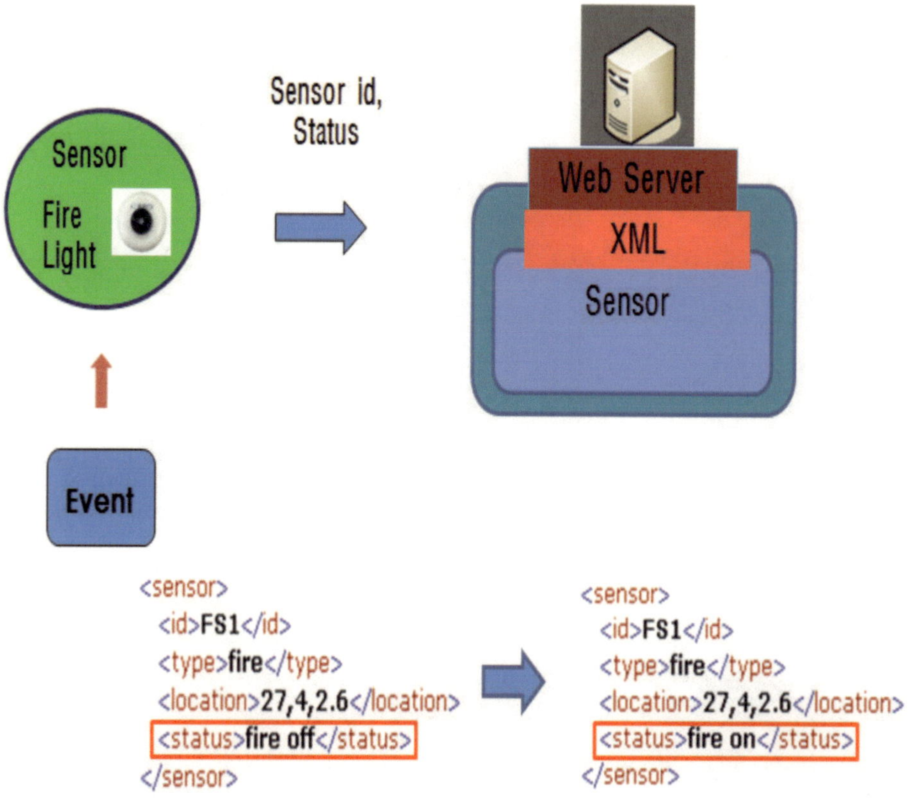

Fig. 15. Automatic robot sensor status update in XML file

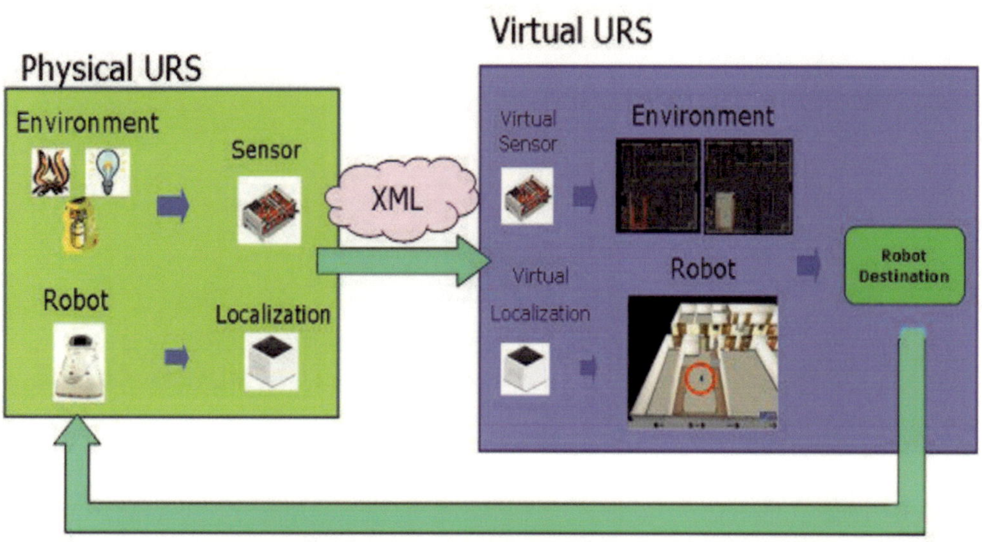

Fig. 16. XML-based bridging between the physical URS and the virtual URS

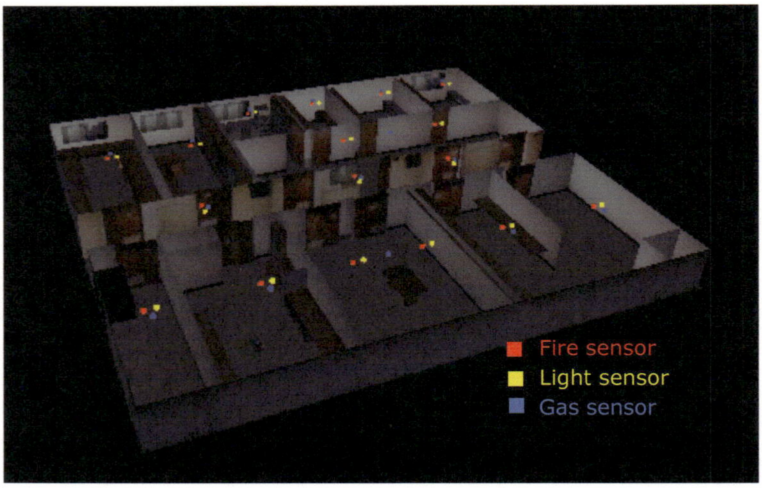

Fig. 17. 3D Responsive virtual URS - 3D visualization of sensor distribution

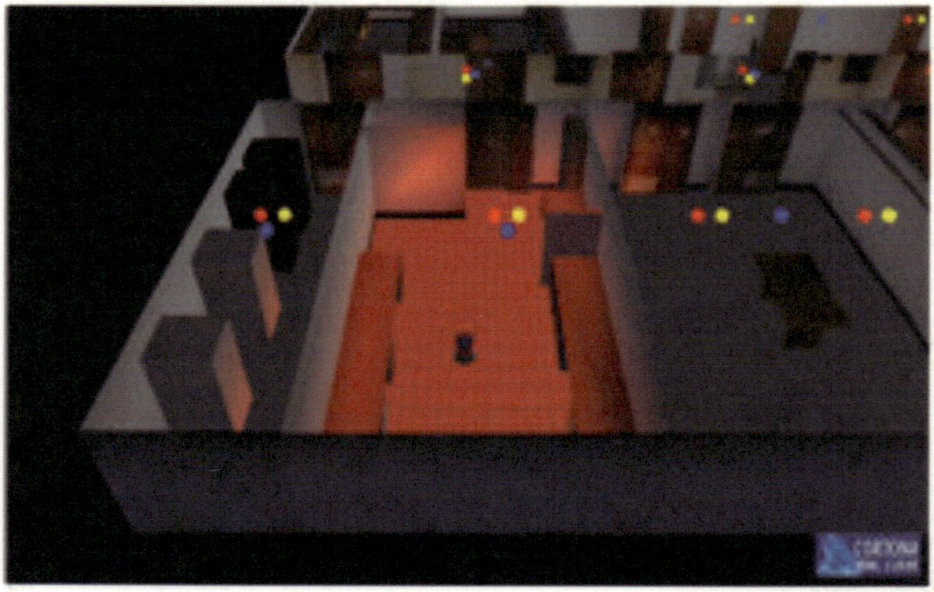

Fig. 18. Fire sensor-based event visualization

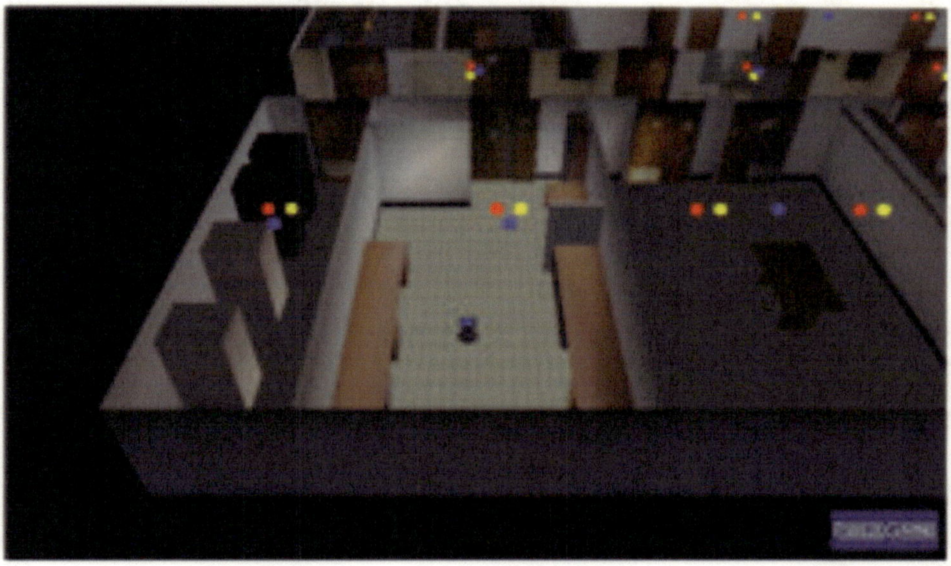

Fig. 19. Light sensor-based visualization

Fig. 20 shows an application scenario of the virtual URS while bridging with physical URS. When fire event occurs, Fig. 20 shows how to coordinate between the physical URS and the virtual URS. Here, the virtual URS visualizes the status of indoor space and a robot will be moved to the fire place for extinguishing fire.

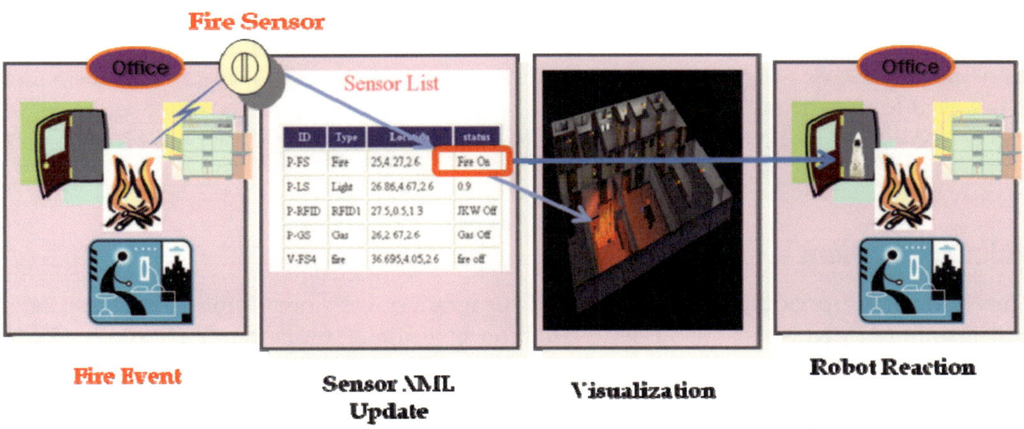

Fig. 20. Application scenario of the virtual URS when fire event occurs

Fig. 21 shows a real implementation of bridging service between the physical URS and the virtual URS. In Fig. 21, when temperature becomes over 50 degree, the virtual URS is responding and the robot moves to the fire place.

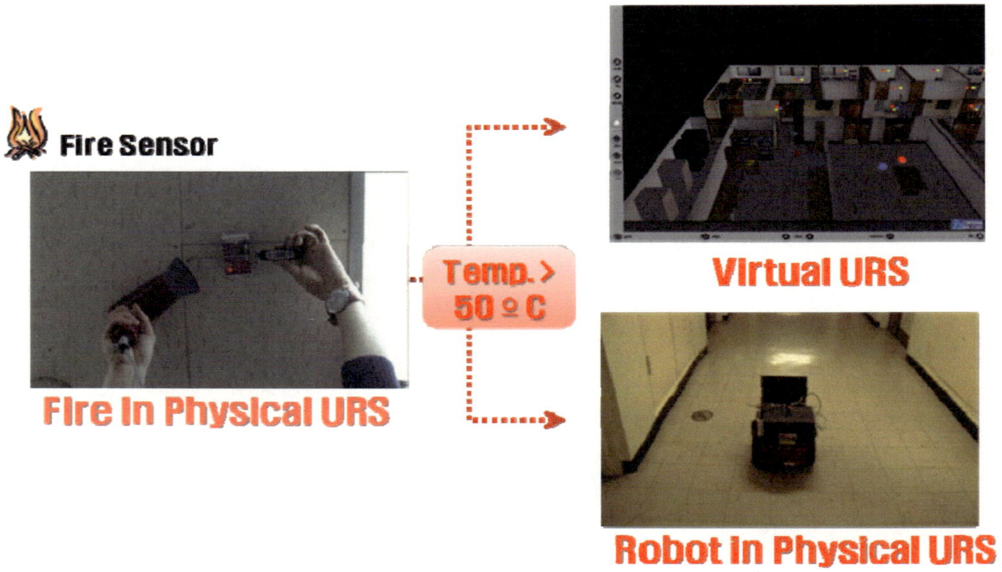

Fig. 21. Implementation of bridging service between the physical URS and the virtual URS

5. Summary

This chapter presents the modeling technique of indoor space and XML-based environment sensor and the robot service technique while bridging between the physical space and the virtual space. This chapter describes our approaches of indoor space and environment sensor modeling. Our sensor modeling system provides sensor XML GUI, sensor XML file generation, zigbee based detection of sensor module and automatic addition of sensor model data into XML file. The bridging system between the physical URS and the virtual URS is also implemented using web server while sensor status is reflected into XML file automatically. Sensors detect the robot position and situation and the detected information is reflected to the virtual URS. This chapter also describes the interactive robot service. User is able to control robot through the virtual URS. The interactive service is possible on mobile phone as well as web.

Acknowledgment

This work was supported in part by the R&D program of the Korea Ministry of Knowledge and Economy (MKE) and the Korea Evaluation Institute of Industrial Technology (KEIT) [2005-S-092-02, USN-based Ubiquitous Robotic Space Technology Development].

6. References

Peter Biber, Henrik Andreasson, Tom Duckett, and Andreas Schilling, et al. (2004), "3D Modeling of Indoor Environments by a Mobile Robot with a Laser Scanner and Panoramic Camera," *IEEE/RSJ International Conference on Intelligent Robots and Systems (IROS 2004)*

Heeseoung Chae, Jaeyeong Lee and Wonpil Yu (2005), "A Localization Sensor Suite for Development of Robotic Location Sensing Network," *(ICURAI 2005)*

Hahnel, W. Burgard, and S. Thrun (July, 2003), "Learning Compact 3D Models of Indoor and Outdoor Environments with a Mobile Robot," Elsevier Science, *Robotics and Autonomous Systems*, Vol. 44, No. 1, pp. 15-27

Kyeong-Won Jeon, Yong-Moo Kwon, Hanseok Ko (2007), Interactive 3D Virtual URS Service based on USN on Mobile Phone," International Conference on Control, Automation and Systems 2007, Oct. 17-20, 2007 in COEX, Seoul, Korea

Y. Liu, R. Emery, D. Chakrabarti, W. Burgard and S. Thrun (2001), "Using EM to Learn 3D Models of Indoor Environments with Mobile Robots", *18th Int'l Conf. on Machine Learning, Williams College*, June 28-July 1, 2001

Wonpil Yu, Jae-Yeong Lee, Young-Guk Ha, Minsu Jang, Joo-Chan Sohn, Yong-Moo Kwon, and Hyo-Sung Ahn (Oct. 2009), "Design and Implementation of a Ubiquitous Robotic Space," IEEE TRANSACTIONS ON AUTOMATION SCIENCE AND ENGINEERING, VOL. 6, NO. 4, pp. 633-640

6

Tele-operation and Human Robots Interactions

Ryad. CHELLALI
Human Robots Medited Interactions Lab. Tele-robotics and Applications Dept
Italian Institute of Technology
Italy

1. Introduction

We are daily and continuously interacting with machines and so-called 'intelligent' manmade entities. We push buttons and we read instructions to get money from cash-dispensers, we tune the washing machine or microwave oven with more or less efforts quasi-every day. Following that, one can easily admit that our era is heavily based on man-machines interactions and the easiness one has in handling such machines is capital, mainly in terms economical, social and psychological impacts. Robots, as a singular sub-set of these machines, are also subject to the same constraints and preoccupations. Moreover and unlikely to mobile phones, PDA or other intelligent devices, interactions with or through robots (tele-operation scheme) are more critical and more specific: interactions with robots are critical because robots are designed to achieve complex tasks within versatile, changing and hazardous environments. They are specific because robots are used instead (sometimes as extensions) of humans (for safety or for economical reasons) leading to confusions between machine-robot and living-robot concepts.
The objective robot (the machine executing a program) and the subjective robot (the anthropomorphic robot and its image in folks mind) are entities too complex to be seen only as simple input-output black boxes. We believe that interactions with and through robots need very advanced and multi-disciplinary methodologies for designing human-robots communication, co-operation and collaboration interfaces.
In this chapter, we give our vision for human-robots interactions. For this purpose, we propose to revisit the robotics timeline. We will show through this timeline the strong relations between robotics and tele-operation. These relations will be depicted under two perspectives: firstly, from *human-robots interactions* point of view and then from *robots autonomy* one. The natural and effective junction between these two visions will take place with the companion robot, e.g. the autonomous robot which is able to co-operate and to collaborate with humans. We belive that before reaching this robotics' ultimate goal, one must answer to a central problem: how humans perceive robots? This formultaion and the answers one can give to the question will undoubtly lead to design effective robots and simplified tools allwoing natural and transparent human-robts intercations.
The document is organized as follow: the first part gives some historical hints letting the reader have a synthetic view of robotics' story. In the second part, we develop our theory about human robots interactions. We will see how we can built a new framework, namely

the anthropomorphic robotics, by combining existing theories coming from neuroscience, psychology and psycho-physics. We show then that this theory can support simple tele-operation problems (delays, cognitive overloads, physical distance, etc.) as well as advanced human-robots co-operation issues.
We finish by presenting some tools we are developing and some exemples of researches we are conucting to assess our hypothesis.

2. A brief Robotics history

In this part we discuss robotics' history. This last has a lot of versions, containing myths, lies and realities. The purpose here is not to establish the exact history; historians will do this work better than us. The idea is to focus on the robotics time line in order to understand what the main motivations in robots development were.

2.1 The imaginary robotics and the pre-robotics era

Robotics historians agree that the first public use of the word ***robot*** was around 1921: it was introduced by the Czech writer Čapek in his R.U.R (Rossum's Universal Robots) play to describe are artificial people. This factual reference came after many other official and unofficial histories of robots or what can be assimilated to robots. Indeed and as far as traces exist, the existence of artificial and human-like beings obeying and executing all humans aims and desires or behaving like them was an essential part of the folk belief. Such mythical characters were largely present and written stories exist for the Greek era (Ulysses et Talos for instance). A more practical idea and a tangible entity were proposed by Ctesibus (270BC). He built a system based on water-clocks with moveable figures. Al Jaziri in the 12th century, proposed a more sophisticated set for the Egyptian emperor: he developed a boat with automatic musician including drummers, a harpist and flautist to entertain the court and the emperor's suite. In Japan during the same period, Konjaku Monogatari shu writings reported a mechanical irrigation doll. These developments were transferred to Europe via Frederic II who received a sophisticated clock from the Egyptian emperor's in 1232. Horology techniques hence received were developed and important new realizations were achieved: Leonardo Da Vinci, for instance, proposed an animated duck in the 16th century and Pascal who built the first computer (*Pascaline* 1645). Jacques de VAUCANSON developed an eating, digesting and defecating duck, which can flap wings also. Many other examples followed during the Enlightenment-era like the 'La Joyeuse de Tympanon' music player offered to the French queen Marie-Antoinette. These efforts were continued and a lot of automaton like chess players, writers, animals, etc was created in Europe thanks to the mechanist stream. This last was not only used extensively to design and build improbable creatures, but also and mainly in industrial applications: De VAUCANSON for instance was also a lot involved in textile industry development in the area of Lyon in FRANCE. show their power through technical capabilities.
Another step was achieved in the 19th century: Frankenstein fiction creature (in 1818) was presented within a movie. Conversely to what was developed before, Frankenstein creation corresponds to a new vision and a new challenge and the movie suggested that humans can create living (in the biological way) entities. One can imagine that the purpose of this movie was to show that humans have enough knowledge to replicate biologically themselves, at

least through their imagination and images and tendency still exists and movies like 'Terminator', 'AI', etc. had great successes the last decade.
In the 30's Asimov emitted his famous rules. Even if real robots did not exist, Asimov had formalized the ethical rules that may govern the relationships between humans and probable robots. His assumptions were purely imaginary and based only on supposed future robots.
The concept of robot perhaps exists since a long time. For sure not having the same meaning as we have it in 2009 but as an imaginary entity able to behave like humans and having an external biologically plausible shape. This entity exists already in the folk's mind that was shaped through mystic and mythological representations in the early times, mechanical during Enlightenment-era, virtual very recently and present today under humanoids or animats umbrella. The other interesting fact is that robots have served as a sign of power, successively mystic, military-industrial and technological.

2.2 Tele-manipulation and Tele-operation to answer to real needs

Since prehistory, humans developed tools to ease fundamental daily life tasks namely, eating, hunting and fighting (*homo- habilis*). To catch a pray or to cook it, humans used very early tools allowing to achieve the previous vital tasks. When considering cooking, humans utilized sticks to avoid to be burned. This behavior can be seen as the first transfer of dexterity at a distance of some cm's and can be considered as the ancestral tele-operation. Closer to us in the 40's, the need of manipulating dangerous products, mainly nuclear substances appeared to be essential for military applications. This leaded to the construction of the first tele-manipulators. R. Goertz and his group developed at ANL a set of prototypes (E1 to E4) of mechanical-based remote manipulators. These researches were done at that time to give operational solutions to immediate and sensitive problems the nuclear industry was facing. The first systems were passive, i.e. tele-manipulators were based on mechanical systems allowing to human forces and efforts to be transmitted to a slave. It is obvious that for these systems both energy and decision making were completely handled by the operator. Thus, one can easily imagine physical and mental operator's heavy workload, leading to a fatigue limiting performances. A first improvement was done by introducing energy into the system. Electrical actuators were used to supply user's forces. sensors and controllers. In such way remotely controlled manipulations were simplified by injecting energy to the system and by discharging operators from low level controls. The further developments of tele-operation were concerned with the introduction of more 'intelligence' within the system. Indeed, thanks to the advances made in computer technology and automatic control theory, some aids were introduced to help the tele-operator and to discharge him from low level tasks. All was done to ease the process to human operators and let them manipulate distantly and dexterously dangerous and toxic products. However, the golden age of tele-operation was supposed to be finished in the beginning of the 60's with the industrial use of the first autonomous manipulators.

2.3 From industrial manipulators to mobile robots

In the 50's and, the industry growth was huge and needs in terms technologies allowing more productivity and lower costs were a priority. Within this context, G. Devol and J. Engelberger decided to create Unimation, the first robots manufacturer. The purpose of the

Unimation robots was to perform spot welding and any other task being hateful to workers. One can notice that these robots were derived from the ANL prototypes: both technological components and morphology are coming from Goertz prototypes. A short way to describe these robots is to consider it as a tele-manipulator where the operator was removed. Somehow, Unimation showed the way for developing robots' autonomy by integrating automation technology. This early introduction of the UNIMATE robot within General Motors and Ford chains was not economically sustainable, its Return On Investment (ROI) was very low compared to the one usually obtained through classical techniques (e.g. man-power). However, the psychological impact on US workers, US customers and more generally on worldwide population was much more than expected. Indeed, as people associates robots with myths and the most technological advances. In cold war and in an economical boom contexts, robots can show to others the USA power and to US customers that their products are perfect.

The next major step for robotics was made in the 70's. The boom of computers and the birth of computer science as an important research topic pushed researchers to look for visible and tangible applications. Indeed, Artificial Intelligence was considered as the ultimate finding and AI techniques were presented as tools able to compete with humans in terms of problem solving [24]. Robot manipulators were used as first demonstrators but it was not enough: the manipulator is equivalent to human hand only, no mobility neither consequent requirement could be tested. To go further and to proof Herbert predictions, the SRI proposed the first autonomous wheeled mobile robot [23]. This last was designed to deal with unknown environment by navigating, avoiding obstacles and recognizing objects of interest. This stream was fully supported by Artificial Intelligence school and served as a proof of AI capabilities [Minsky, McArthy]. For robotics community, these developments can be interpreted as the first efforts in increasing robots mobility, .i.e., transforming the manipulators into mobile platforms to face a wider world. This additional mobility the necessary integration of new aspect of robotics such as perception and locomotion. By doing so, e.g., by developing autonomy and locomotion, it was possible for robots to enlarge their working space from the assembly cell to wider and more complex environments.

Unfortunately, 30 years later the problem of autonomy was unsolved. [2] with his famous 'Elephants do not play chess' and many others [26] gave some explanations to AI failure. Nevertheless, many major advances have been achieved. Actuation and mobility were largely visited (mainly automatic control and sensors technologies) and very efficient solutions are existing actually. Indeed, efficient mechanical structures were built, from small bugs to humanoids to address locomotion. Moreover, interesting solutions were proposed letting robots walk, fly and swim with high accuracy, large mobility and stability. As well, computational and sensing capabilities have been also improved to a fascinating level. These technological solutions enabled to handle many aspects, mainly, low level controls which could be considered as solved. Unfortunately and despite all efforts and technological developments, decision making and autonomy are remaining bottlenecks.

2.4 Humanoid robotics

In parallel to the SRI-SHAKEY project, the University of Waseda launched the project Wabot-1 (1970). The aim of the project was to develop the first full scale human-like robot. It was presented in 1973 as the first anthropomorphic robot. This was a major jump for robotics: the idea behind was not to serve only as a test-bed for Artificial Intelligence developments but also to focus on its main property: the anthropomorphism. This last characteristic or property was considered as the necessary condition for robots to be *accepted* as companions. Indeed, the Wabot-1 was targeting the field of human assistant or human companion. It appeared early that the necessary condition to achieve such a goal was its human-like shape: the robot needed to be designed as a human. This morphology was supposed to be the logical way to allow the robot to evolve in the same indoors environments and to satisfy all the consequent (i.e. man made environments) constraints. Another motivation for shaping the robot this way was perhaps the facilitation of the human-robot interactions. The last motivation is a pure and personal speculative interpretation: by creating such a robot, the developers were targeting high emotional, psychological and economical impacts, e.g. it was a way to show that the Japanese industry is on his way to master and to reach the top edge of technology as it was done earlier in USA with manipulators.

The humanoid robotics was there and researchers faced more complex problems than for wheeled platforms. In all sectors, the humanoid presented the most challenging issues: for motion control, for environment understanding and perception, for interactions with humans, etc. the humanoid requested more advanced solutions. One of the first efforts was dealing with locomotion. This issue is specific to legged robots and conversely to wheeled robots, legged robots in general and bipeds in particular need a dynamic stabilization while walking. This inherent issue was addressed early and the ZMP [4] formulation was proposed and many automatic control based solutions were proposed and implemented within this framework. More recently, bio-inspired approaches (CPG's for instance [27]) were proposed and implemented. From the mechanical point of view, specific solutions were developed to ease the stabilization, mainly bio-inspired structures and parts (including compliances, stiffness, etc.).

Perception was also a big issue and one of the most challenging topics. The first proposed solutions were directly derived from classical approaches. Computer vision-based techniques for instance were applied to navigation and object recognition. Likely, reasoning and cognitive researches adapted existing ones and transferred to the humanoid context.

Unfortunately these efforts were not sufficient and extra-robotics help was needed: the humanoid theme is considered nowadays as a transversal project where the cooperation of many disciplines (like neurosciences, bio-mechanics, nano-bio technologies, sociology, philosophy, psychology, etc) and fields in mandatory. Following that, important transversal initiatives through labs and institutions were launched [Icube project, HRP project, COG project, etc.].

However, we can notice the existence of two main streams targeting two different goals exist: the first one aim is to develop co-operative robots, e.g., having capabilities and abilities to understand human's desires, behavior, speech, etc having emotions and proper behavior, etc. The second stream uses humanoids as test-bed to better understand humans: the humanoid is used as a simulator to support human-based models assessments [25].

Humanoid robotics is in its early stages and the current work within the two previous streams can help us to learn more about us in order to derive therapies. Rehabilitation techniques and other prosthesis are part of the goals of humanoid robotics. It can help us also to design better interfaces in order to simplify and to ease the use of our daily life machines and tools [28].

2.5 What can we learn from the past?

From the previous brief history of robotics, one can derive some conclusions and some lessons. This may help to better understand the current status of robotics and to identify the targets of future researches.

Fig. 1. Robotics timeline

The first observation one can make is concerning the singularity of the entity "robot". This last is a singularity and it is perceived as a particular entity; not an amorphous one and not a living one as well. This leads humans to develop very specific relationship with robots and there are differences between reactions people have when they interact with usual machines (washing machines, mobile phones, etc) and robots. Fundamentally these differences are small (except shapes and nature of functions they perform): both are materiel, they accept controls and orders execute programmed tasks and give feedbacks about their status. Does classifying a machine as a 'robot' changes its status? Does the shape play a role in this classification? And how this machine is then perceived by humans? The robot generates a lot of open questions which are still under investigation.

Some subjective explanations may be found in the pre-robotics era. Humanity belief at that time was largely influenced and shaped by myths of super manmade creatures. Later, the human-like automaton was no more a myth but a fascinating and tangible human creation. This fascination has probably created a specific status for these inanimate human-like entities and thus prepared the current perception of robots.

This last is subject these days to a lot of research and many efforts are done to qualify and to quantify this specific status. In other words, researchers are trying to determine how humans (and animals also) perceive robots. Social and emotional robotics shows that this perception is not unique and a lot of parameters are taken into account. Studies concerning robots' design and shape, embodiments in terms of animacy and intelligence, the age of users, etc have effects and impacts on human-robot interactions. One can imagine easily how the results can be used in the frame of companion and co-operative robotics.

However, these studies and the resulting hypotheses/theories may be moderated for tele-operated robots. Indeed, one may be aware that these last and autonomous robots have a fundamental difference: tele-operated robots have to achieve crucial, 'zero error' and measurable tasks while autonomous robots have more freedom and their errors are tolerated. Following that, the relationships between humans and the two categories of robots (e.g. autonomous and tele-operated) are different; in goal directed and 'zero error' interactions, human must adapt and must compensate the remote robot's limitations. For autonomous robots, the interaction is less constrained because the user is somehow less demanding.

The third observation is more technical and concerns the modern robotics evolution. It makes sense to consider the first tele-operation systems as the ancestor of modern robots. As we said before, the robots' evolution was pushing developments and innovations toward autonomy. The timeline started with human-robot systems to move to systems where more 'intelligence' was included. Consequently, human presence was minored. However, robotics returned to root and to tele-operation whenever an operational system was needed: the presence of humans in the control loop was considered as the ultimate and safe solution.
The last observation is concerned with the contributions of the current tele-operation. This last is still improving human-robots interaction field. Tele-operation systems are the only ones putting in close and constrained contact robots and humans. This forced synergy pushes operators to adapt to machines, pushes engineers to find the best interfacing technologies and pushes researchers to understand human to build effective systems able achieve safely critical and vital tasks.

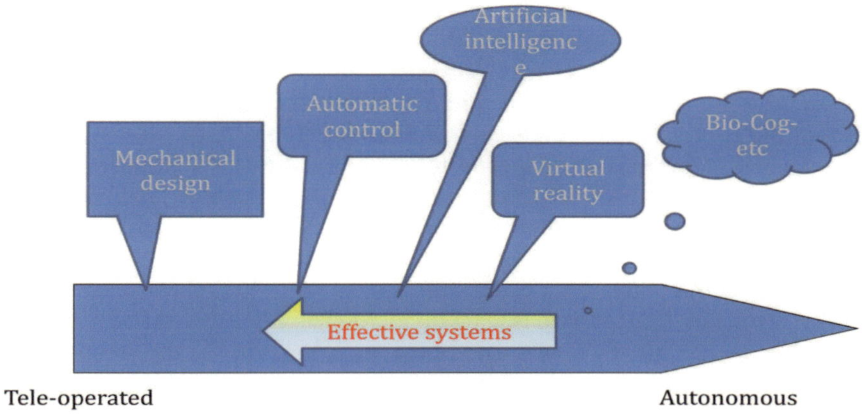

Fig. 2. contributions to robotics

3. Tele-operation as a starting point to built a new vision of robotics

In this part we propose to describe the current tele-operation systems. This will allow us to highlight the most important achievements, the advantages these systems and to finish, we will try to list their drawbacks and we discuss these last.

3.1 Current Tele-operation systems

First tele-operation objectives were to enable to control distantly complex mechanical systems, in order to (tele) manipulate dangerous and hazardous material. The first need at that time was concerned with dexterity: the system has to be enough dexterous to support simple tasks to pour some liquid into a bowl for instance. Four degrees of freedom mechanism (a pincer with a spherical link) was proposed to handle this class of tasks. Rapidly and after dexterity, other needs came-up. For safety reasons, the distance between the operator and the operation field had to be increased. This leaded to new needs, namely, a complex mechanical transmission was implemented. The first master-slave system developed by Goertz was purely mechanical and operators moved the slave through wires and tapes while having a direct view of the slave. By doing so, operators handled parasitic inertia and thus they produce additional workload not directly dedicated to manipulations. This distortion was corrected soon after. Indeed, electro-servomechanisms and controllers replaced wires and tapes. This allowed an electrical force reflecting position letting operators dedicate their energy to manipulations only.

After mobility, dexterity and force reflection improvements, people moved to ameliorate other sensing capabilities. Indeed and to protect users, a minimal distance between the master and the slave was imposed. Two problems rose with this mechanical separation: direct vision was no more possible and time delays appeared. For the visual feedback, simple live video streams were displayed on simple TV screens and for delays, people started by trying to understand its effects [11], namely they performed the first psychophysics studies to model human behaviour when performing tele-operation tasks. Their conclusions were that operators utilize the 'move and wait' strategy in presence of delay; humans compensate internally the closed loop delays. This leaded to the development of the so-called supervisory control; a heuristic approach where the controller allows the operator to specify tasks at a high-level. These tasks are decomposed by the controller into atomic commands and performed by the remote controller as a suite of simple tasks. The sequence is executed under the operator's supervision (.e.g. the operator can interrupt the process at any time; he can also modify the task's description content or level). This symbolic (or AI based) approach leaded to software solutions to provide more 'intelligence' and autonomy to the remote controller to compensate delays. A variation of this approach was proposed later with the notion of predictive displays [10].

This approach was followed by a huge effort from the automatic control community [29]. For the latter, tele-operation has been stated as a two folded problem: stability and transparency. For the first aspect, the goal is to maintain the stability of the system regardless of the behaviour of the operator or the environment. For the second, the goal is to allow tele-presence feeling, .e.g., hide the interface and let the operator perform interactions as he was within the remote environment. Many advances, mainly Lyapounov based analysis, impedance and hybrid representations, passivity based control schemes, etc., were made allowing stable and very efficient solutions to handle inherent delays like in space, underwater applications. Likely, transparency was tackled through the two ways transmission of force and velocity.

Nevertheless, force reflecting and visual feedback appeared very early as insufficient sensory modalities to guaranty efficient remote interactions: operators need more than 2D viewing and haptics-based links with the remote world. More sensing technologies and displaying devices were integrated or developed to improve existing systems in terms of immersion [11] [12], [17].

A lot of work has been done for instance on the visual channel. Sheridan summarized the influence of video feedback on tele-operator performances. Frame rate effects, resolution, colors, occlusions and position of the operator's point of view were also studied. It was shown that performances were affected. Haptics channel received also a lot of attention. Force feedback arms used are the typical and the most studied bilateral ways of controlling slave robots: it closes the loops of force-torque based interactions.

Many similar technical solutions were proposed for other channel, mainly tactile, auditory kinaesthetic and even olfactory. The guidelines for the design of such tools were directed toward reproducing as exactly as possible information, actions and reaction flows for both sides (the master and the slave): on one hand, the master (human operator's side) must capture all the desires of the human operator. On the other hand the slave must capture the "image" of the remote world and translate it into stimuli for the operator to let him fill itself within this world. This latter concept or definition is well-known as tele-presence. It was introduced in the late 80's by S. Tachi [20] to describe systems allowing users to feel them self within remote worlds. Asymptotically, tele-presence systems are the ones enabling to humans operator to tele-exist, i.e., not only to feel being somewhere but also letting distant people feel the presence of the tele-existent person. In fact both systems are theoretical models and their practical implementation is limited because of technical and fundamental barriers.

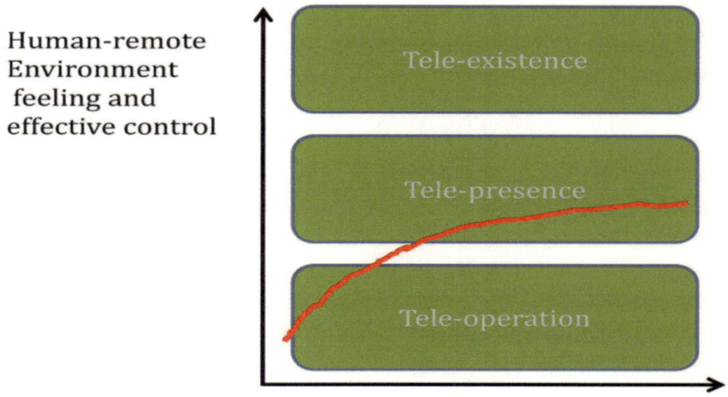

Fig. 3. toward tele-exitense

In parallel to the previously described technical efforts, some works tried to reconsider the tele-operation problem from the human-centred system point of view. Indeed, human is a central piece of master-slave systems: he issue commands function of what he perceives from the remote world. Following that, ergonomics and human factors appeared within tele-operation field and several studies were conducted. These latter were initially inspired and derived by previous works in man-machine and man-computer interfaces. The problem was stated as the handling of complex systems and sensory feedbacks and input devices were the identified key issues. The formulation was the following: to let people interact distantly, one needs to collect the maximum information about the remote world and display this information as soon as possible and as accurate as possible to operator. Likely, to transmit orders and controls, efforts were made to construct simplified and effective input tools.

3.2 Current systems drawbacks

In literature, tele-operation systems drawbacks are mainly identified as consequences of distortions or/and the absence of sensory feedbacks or/and the weak knowledge of the slave. This is partially true. The part of truth is due to technological limitations. It is easy to notice that current sensing, transduction and displaying technologies cannot reproduce stimuli (at all or at least partially) nor capture intentions that should be generated directly and not synthetically (e.g. through interfaces).

Indeed, the current technology have access and can measure only few human parameters like gestures, speech, forces, torques, postures, direction of sight, etc.... Likely, to display the remote environment, current systems transmit incomplete and distorted information, like live video flows, forces, torques, sounds, etc... In both directions, information is partly missed or distorted. When controlling the slave, the operator compensates this lack of information for both sensory and motor aspects. He rebuilds mentally the remote space from the available feedback fragments. As well, he generates, the right slave's controls through the mental representation he has about the remote system. Said differently, the operator try

- To build a perfect matching between his space and the slave's one by compensating missing parts and by removing all the sensing and displaying based distortions.
- To build a model of the slave

The previous points could explain the cognitive overload situations and operators fatigue arising when remote interventions are performed or more generally when a human is piloting a complex system. This conclusion has to be reconsidered for tele-operation systems.

The other issue is more fundamental and it is concerned with unquantifiable parameters. Indeed, within tele-operation systems, the slave robot is a constructed as a machine with a very specific set of capabilities. However, people perceive it in a dual way: it can be seen as the classical tool that one can use to modify the environment or it can be seen as the semi-autonomous entity obeying and accomplishing human commands and desires. In the first case the tool is considered as the operator's body extension, i.e., a mean to increase the personal working space. In the second case the slave is considered as an exogenous entity supporting human orders and informing him about its status and its environment. The two schemes and visions have different implications on the operator's sensory-motor system. On one side, the slave robot has to be integrated within the operator's sensory-motor space as part of his body. On the other side, one needs to build a mapping between two heterogeneous sensory-motor spaces.

Nevertheless, the two highlight the core problem, namely the existence of fundamental differences between the operator and the slave. These differences are the following:

- The difference of dimensionality between the operator's and the slave's sensory-motor spaces,
- The differences between stimuli perceived in direct interactions and the ones synthesized by the system's interfaces (a direct view and an HMD based display is a good illustration).
- The differences between the operator's physical actions (on the interface in this case) and the ones really achieved by the slave, e.g., the physical modifications of the remote world.

The previous differences reflect the distances tele-operation systems designers are trying to reduce. It overpasses the sole Euclidian space with his physical distances, time delays, scale changes, etc to include sensorial and motor spaces to create the optimum matching between the human and the robot spaces.

This formulation is just a new way to express the goals tele-operation systems' designers are aiming to reach, namely to reduce these differences or asymptotically to have the perfect tele-existence system. The main issues are still there because the human sensory-motor

space is hard to describe and thus the metrics needed to operate on this space do not exist yet. People use intermediate spaces and indirect measurements, mainly derived from psychophysics, ergonomics and human factors, to assess or to design tele-operation systems. As said before, we have two working hypothesis: the system is considered as a body extension or as a semi-autonomous entity. Following the one or the other hypothesis, one needs to adopt a specific methodology to reduce the differences between the two sensory-motor spaces. In other words, the knowledge the operator must have and/or acquire about the slave is different. In the first case, the slave has to be integrated implicitly within the operator's sensory-motor system (e.g. considered as prosthesis). In the second one, the slave is considered as a collaborator with limited capabilities. Thus it requires the generation of specific controls and the development of specialized understanding skills. In both cases we can find a *humanization* aspect of the machine. This is specific to tele-operation and absent in other classical remotely controlled systems.
The last point, but not the least, is concerning the operator's sensory motor space. As presented before, this last is appearing like a classical vector space with linearly independent vectors set as basis (each corresponding to a sense or to a motor activity). This representation is missing an essential part, namely the cross-relations and the couplings between the sensory and the motoric components. In a hand-eye based action like catching for instance, any defect in one or other part influences greatly the other one. Researches taking into account the coupling effects started some years ago and they confirm the importance to reconsider the sensory-motor space construction and its use in the design of tele-operation systems. As a consequence of the previous finding is the following rule: *motoric anthropomorphism is necessary but not sufficient to have an effective tele-operation system.* An exoskeleton for instance cannot guaranty the efficiency of hand based interactions and other percepts (visual, tactile, kinesthetic, etc.) are needed.

To conclude and to open the next section, we can speculate on the ideal tele-operated system as oneself person: if one has its own image as a slave so he will make no efforts in controlling it and performing any kind of remote world transformation. The bijection between the remote and the master space and the mapping are perfect and no extra-efforts are needed to execute remote tasks. Somehow this is the asymptotic goal of tele-operation and amazingly humanoid robotics.

3.3 Toward robot's autonomy and the necessity to have humans within the control loop

Robots autonomy is one the first dreams of robotics' research. The pending and central question of robotics is the following: how to make a machine which is self-sustainable, e.g. able to move safely, to find its own energy, able to understand and to communicate with humans and other robots, etc. Many other capabilities can be added to this open list; dexterity in manipulating objects, recognizing these objects, understanding contexts and situations, etc.
This dream can be heavily moderated when having a look to autonomy definition or definitions. Indeed, we found plethory of definitions, each suggesting a singular and domain-dependent point of view. The most generic one is the following: "giving oneself' own laws". As expected, for robots and robotics, this definition is not fully true. Indeed, people program robots. By doing so, they transfer parts of their knowledge to robots. This knowledge is derived from the thoughts of the programmer and it reflects his answers to

specific conditions (the task requirements). In other words, if a robot has to face a task, the programmer will imagine all the possible situations and consequently, all the suitable solutions to achieve the desired tasks. Certainly, learning, development and evolution procedures can increase robot's degree of autonomy, but formally, robots programming is equivalent to *a priori* tele-operation. One imagine a situation or a goal, derive the consequent robot behavior and the program it. This could lead the illusive autonomy and the robot will fail when facing unseen or unknown situations. This phenomenon can be illustrated through a parabola: the robot is put in a tunnel and the only way for him to evolve is to go back and forth without any chance to leave the tunnel, e.g. no way for the robot to find out an alternative to the imposed pathway. The robot's behavior is thus predictable and this is in contradiction with autonomy definition. Obstacle avoidance task is a good example for what I call *illusive autonomy*. Indeed, at a first glance, all obstacle avoidance behaviors implemetations are fascinating and could be considered as intelligent behaviors. In fact, the statement for this class of problems is mainly a measurement-based: robot sbuilds the geometry or the topology of the surroundings and adopt very simple algorithms to find out a free path.

Many other problems could be seen analogously as sensing-measurement problems (object recognition, localization, etc…) rather than advanced behaviors and real autonomy. Illusive autonomy is a consequence of designing biological like behaviors, acceptable for observers but without any justified foundations.

It appears that programming robots, namely transferring knowledge to robots is one of the key issues for building autonomous robots. We transfer methods and procedures, namely logical suites of actions hoping that it will cover all situations. One imposes partial predefined behavior and mechanisms to allow robots to handle situations we suppose them to be and to face.

The main question arising thus is the following: how to make such mechanism generic, e.g. the robot can learn new behavior without programming? As for children, robot cannot learn without the help of a more experienced entity (human or robot). The learning process needs examples and more than that, living examples.

Two sub-questions arise then: do we have the right hardware to support such mechanism and how to let the robot understand examples given by a more experienced entity.

The first sub-question is itself a research area and we will not address it here. Indeed, regarding hardware some functions are implementable others not: one cannot overpass the potential capabilities. Humans have a genetic potential leading to advanced behavior like adaptation. Animals for instance cannot overpass certain barriers: an herbivore cannot eat meat and become a carnivore even if its life is in danger. Changing alimentary regime is impossible (at least for short term horizons). A fish cannot run on the grass while humans can both swim and run on the grass. They adapt to learn swimming and more complex, they create specific tools to change their nature to go for instance underwater.

The second one leads to reconsider the human/robot robot interaction under the learning-transferring knowledge point of view. Tele-operation as the main field putting humans and robots together to achieve physical interactions may be a good candidate toward autonomous robots. On the other hand, if we assume that we have already autonomous robots, these last are supposed to interact with humans. Here also, a revisited tele-operation may play a major role to facilitate humans-robots communication [robonaut and tanie].

In this part, we proposed a new point of view from which tele-operation may be seen. In addition to be the tool of modifying physically remote, distant or inaccessible worlds, tele-operation also:

- Can help to design the right interaction paradigm between robots and humans,
- Could be an alternative solution to support the design of autonomous robots,
- Could be a tool to better understand humans.

4. The anthropomorphic robotics for a new formulation of tele-operation

The mechanical anthropomorphism introduced as the necessary but not sufficient condition to simplify the human robot communication and control. It simplifies the matching process between the human and the robot motor sub-spaces and thus allowing effortless control. The anthropomorphism I want to introduce here is a generalization and concerns the whole sensory-motor system. This generalization is purely speculative but it can be supported by a strong background and can be used as a framework to design efficient tele-operation systems and more generally, efficient interfaces. To do so, I rely on two existing findings in psychology and neurophysiology fields:

- The empathy and more specifically the *perspective taking* theory,
- The theory of mind and his neurological substrate, the mirror neurons.

4.1 The empathy and the perspective taking

1) *The empathy*

The concept of empathy appeared at the end of the 19th century within German philosophical circles. It was concerned mainly with human ability to "feel into" nature and man-made objects and the underlying question about the understanding of human aesthetic objects' appreciation. The central problem was to know why we perceive beautiful and ugly objects and how we use and sense data for our investigation of the world. Lipps extended the concept in the beginning of the 20th century to overpass the aesthetic area. He claimed that empathy should be understood as the *primary* mean for our perception of other persons as minded creatures. At that time, this concept was the only alternative for conceiving of knowledge of other minds. It was described as a process with three steps that enable one to attribute mental states to other persons based on the observation of their physical behavior and one direct experience of mental states from the first person perspective.

- Another person *X* manifests behavior of type *B*.
- In my own case, behavior of type *B* is caused by mental state of type *M*.
- Since my and *X*'s outward behavior of type *B* is similar, it has to have similar inner mental causes. (It is thus assumed that I and the other persons are psychologically similar in the relevant sense.)

Therefore, the other person's behavior (*X*'s behavior) is caused by a mental state of type M. This inference mechanism was largely used to explain social behavior of humans and the way they establish relationships. Nevertheless, this stream was criticized and abandoned to the theory theory approach (theory of knowledge acquisition, developmental phenomenon, learning mechanisms, etc.) which found his applications through AI. Empathy was considered as a very extremely naïve conception of human sciences to explain social relations, the influence of cultural context in human-human understanding, etc.

For our purpose, empathy through the findings, the tools and the methodologies developed around this question in various areas like the human sciences, philosophy and more recently in neurosciences can be a good framework for tele-operation systems improvements. In other words, if the hypothesis of robots' and tele-robots' humanization is true, then human-human interactions knowledge can potentially transferred human robots interactions and recent works tends to demonstrate objectively the validity of this approach at least for humanoids[Krach].

4.2 The perspective-taking

More than for empathy, there no exact or unique definition of perspective-taking: it is research field dependant. If we consider for our needs and our purposes, we will consider its materialist side and somehow a geometrical point of view of empathy. We can define it as the ability one has to drift in and out of his point of view and how this drifting leads to the building of the so-called 'god's eyes view' [30]; *If I'm at someone else place then I can feel what he feels and thus I can understand him.*

Indeed, in one of its versions, the perspective taking theory was concerned with spatial cognition [31]. Following Berthoz [19], the *spatialialization* or perspective-taking allows one shift from a world's egocentric point to view to an allocentric one. This process is considered as an essential process and one the main components of empathy: it materializes and it describes objectively the way we can take others points of view at least to experience their surroundings.

For tele-operation the consequence is immediate: do tele-operators project themselves on the remote robot and construct a remote point of view to achieve physical interactions? A lot of experiments concerning this topic are in progress and partial and indirect answers to this question through experiments are already found. However, is still an open question to be developed in the next few years.

4.3 The mirror neurons and neurosciences

Nowadays Empathy is back. The first revival occurred in the 80's with the simulation theory. This theory conceives ordinary mindreading as an egocentric method where one uses itself as a model to simulate other people's state. More recently and thanks to important findings of neurosciences, empathy can be again considered as an interesting framework to explain how a human recognizes another person's emotional states and intentions. Indeed, empathy received some empirical confirmation through mirror neurons. Neuroscientists have shown that there is a significant overlap between neural areas that underlie our observation of another person's action and areas that are stimulated when we execute the very same action [32]. In other words, they discovered that same brain areas are activated both when a motor activity is observed and performed. Likely, they showed also that humans simulate the motor activity within the mirror neurons area before executing it. This last argument is to add to those of the simulation theory defenders and as a contribution to the rehabilitation of empathy at least as a sustainable framework to explain low level interactions-relations take place.

The empathy together with mirror neurons research is very active. A lot of issues are still pending and no evident proof has been found to explain in detail the underlying processes. However, a lot of other areas are taking advantage of this formalism and apply it to several domains mainly in psychotherapy, education, art, business and economy, etc. Recently, some researchers investigated the extension of empathy to non-human beings. The question was to see whether or not humans can develop empathy toward animals, namely pets. The results are very surprising and might open new perspectives. Indeed, we just have formalism and some experimental evidences. The exact mechanism is not well known and practical and conceptual questions like the following are still open:

1- Can we have empathy with other biological systems?,
2- What information can human extract from seeing partial information about other humans?,
3- Do we interact better if we are manipulating something equivalent to biological systems?

The previous section introduced very briefly a framework which could be very interesting to our purpose. Indeed, the natural question that one could have is concerned with the transfer of the empathy framework to non-minded creatures like robots. We have many ingredients like "simulation theory", "perspective taking", "projection in other's mind" letting some freedom to speculate and answer affirmatively.

4.4 Does empathy with robots make sense?

By essence, the empathy with robots is hard to define and hard to obtain. One is in presence of non-minded and non-biological entities. But before going further, let me tell you what the reaction I had with humanoid robots was. Years ago, a colleague of mine showed me his humanoid robots team playing soccer during a RoboCup contest. After some minutes, I laughed for no apparent reasons. I saw the video twice and I had the same reaction. The way the robots were kicking the ball reminded me my childhood with children (including myself) performing the same actions with naïve and exaggerated imperfections. Was my reaction strange? Anyway, it was for me a questioning situation. A similar situation occurred some months later when I presented a humanoid robot in an elementary school: I observed strange and also questioning reactions. The perception of humanoids and robots in general is a key issue that must be well understood. Basically, we use and we perceive unconsciously some cues and some features leading us to construct an image of the robot. The uncanny valley phenomenon is certainly one of the first experiments trying to elucidate human robot's perception. Solving this issue is fundamental because it may allow simplifying the interactions between robots and humans. Consequently and in the light of we said before, the framework of empathy can be valid in this quest.

For us, the approach may be gradual: one needs first to start with tele-operation systems. Once these systems understood, one can move and tentatively generalize the findings to autonomous systems. For tele-operation systems, the empathy framework can be used as a basis to perform experimental researches. Indeed, for those considered as body extensions and those considered as exogenous entities, the integration within the operator's body scheme respectively the simulation scheme (e.g. operators simulate the motor-activity before sending the corresponding controls to the slave) can be applied.

In addition to offering a well developed set of experimental procedures and methodologies, the approach we are proposing can be evaluated objectively. Indeed, one of the interests of the empathy is that it shows where to search the effects and how to measure it objectively. It is obvious that neither the brain model nor the interpretation of its signals are available these days, but it is the only way to measure direct effects and thus to avoid classical indirect psychophysics-based interpretations.

5. Virtual Reality as a major tool for anthropomorphic robots assessment

Virtual reality is nowadays a major component of tele-operation systems. The acquaintance and cross-fertilizations between the two are old [Coiffet]. More specifically, VR is largely used for both simulation and remote control. Indeed, in its simulation side, VR (and augmented reality) allows to operators to experience interactions with synthetic worlds by synthesizing stimulations (obtained from simulations of real sensors performing under real physical laws) for our main sensory channels like vision, auditory, tactile, haptic, kinesthetic, etc…

In its remote control version, the use VR/AR techniques is mainly dealing with sensory compensations, corrections or substitutions. Respectively, VR/AR systems recreate missing information, remove distortions and enhance sensory signals or perform transfers between senses (visual information is displayed as a tactile one for instance). That is said, VR/AR can be considered as a very flexible stimuli generator and one can address sensory channels with a wide variety of stimulations. This maturity is partly due to robotics and to tele-operation and their strong needs of operational and robust systems. This obliged VR people to find out new interfacing solutions covering the large spectrum of senses with high reliability and robustness.

Naturally and regarding the possibilities VR is offering, human centered researches (like cognitive sciences, psychology, neurosciences, etc) came to VR. These technologies are enough flexible and enough powerful to enable to these new demanding fields to setup new approaches and new experiments for understanding the human brain, its functions and the way it process/analyze external stimulations to built complex and powerful behavior. These systems can support infinity of scenarios addressing *all* human senses in a cost-effective way.

Following that, VR is the best candidate to support the assessment of the empathy-based framework and its translation into the anthropomorphic robotics hypothesis. Following that, we started to perform some experiments but we rapidly faced an unexpected problem: VR is a set of tools and not a science. In our case, a problem concerning depth perception raised when we performed experiments dealing with interactions within unfamiliar environments (with unfamiliar physical laws like nano-spaces or micro-gravity spaces). We obtained biased results even if our visual displays were very well calibrated geometrically. The visual perception and specially depth perception within VR systems is still an open problem to be solved or at least to be well known to avoid biased use and incorrect analysis. For other senses the same conclusion is valid and especially when different modalities are combined.

5.1 Is VR fully reliable: an example through depth perception

VR is supposed to recreate real worlds by stimulating human senses according to natural laws. However, VR cannot generate all the possible stimulations one can perceive. On the other hand, for those possible, the stimulations are mostly distorted or incomplete. Following that, the use of VR is not as magic as it is said. One has to take care with it and to understand all the undesired phenomena VR can induce directly or indirectly. In the next lines, a specific and known problem is given to illustrate the fact that VR has to be more effective.

Immersive viewing devices are key elements for virtual reality systems. They address the richest sensory channel (supporting 80% of human sensory inputs) and thus, regarding the rendering quality, users may experience more or less realistic environments and interactions with these environments. Unfortunately, the above mentioned quality is depending on parameters which are not well understood. Namely, displaying devices affect both perception and actions in virtual environments in a hidden way. This leads to malfunctions and biases in sensitive applications like psychology research and therapy, training or education. Undoubtedly, absolute distance or depth perception is one the main issues and one of the most investigated topic in VR. Many research efforts have been performed to determine the effectiveness of different cues necessary to perceive depth.

For instance, many research reported a systematic underestimation of distances when HMD's are used compared to the same estimation in the real world. Many hypotheses were emitted to explain this phenomenon:

- the reduction of the field view effect,
- the weight of the HMD effect,
- the difference between the viewed world and the experimental place,
- etc.

Indeed, several studies on distance perception using the HMD have found significant underestimation of egocentric distance, the distance from an observer to a target. It was shown in that this underestimation is not due to the limited field of view of a user while using the HMD. In [18] for instance, it is argued that mechanical properties can play a role in the underestimation phenomena. Other explanations have been also given like lack of graphical based-realism or mismatches between the viewed world and the experimental environment (e.g. subjects are aware that the viewed scene does not correspond to the place where the experiment is performed). Likely, other works showed that other sources like visual cues (such as accommodation and convergence) or situations (visually directed actions) may also affect distance or depth estimations and thus decrease the feeling of immersion. In sum, the identification of sources leading to the distance misestimating effects for HMD's is still an open question. We verified the lacks found in literature. In our work we aim to verify the above mentioned phenomena. Our approach is based on the comparison of performances between HMD's and stereoscopic wide screens when simple verbal and relative depth estimation is achieved by seated subjects. By doing so, we simplify the experimental conditions and we concentrate only on few variables. Namely:

5.2 Some examples of VR use in the context of tele-operation systems enhancement

Hereafter, I give some examples of what VR can do. These examples are parts of our project dealing with tele-operation. The first one describes the experimental setup we are using to assess empathy with robots. The second one is an illustration of the possible derivations tele-operation can have.

2) *Empathy measurement: a tentative experiment*

Our goal with this experiment is to verify the hypothesis concerning the existence of an empathy-based relation between human and robots.

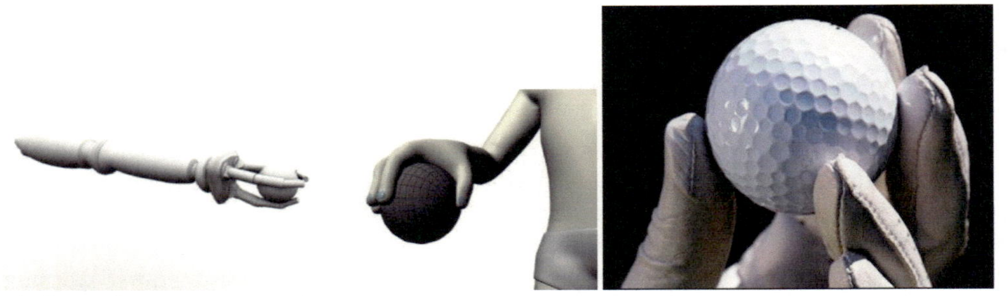

Fig. 4. Empathy measurements

The setup we built is based on a set five hands. Four of them are synthetic with respectively 3, 4, 5 and 7 fingers. The last one is realistic (a copy of a human hand). All of them are controlled by users through a data glove. We will not discuss here the experiment and the preliminary results we have but just to highlight the usefulness of VR in terms of flexibility: it allows us to enlarge our spectrum of potential stimulations to see how users project themselves on anthropomorphic and non-anthropomorphic structures including funny and unrealistic targets (a hand with 7 fingers).

3) Sensory substitution as a minimalist way to improve interactions

The second example is dealing with sensory substitution. The idea behind is to see how can we replace a sensory modality by another one. Namely, we want to study the effects of displaying geometrical information for navigation purposes through the tactile modality. One can imagine easily that for blind people such substitution is important.

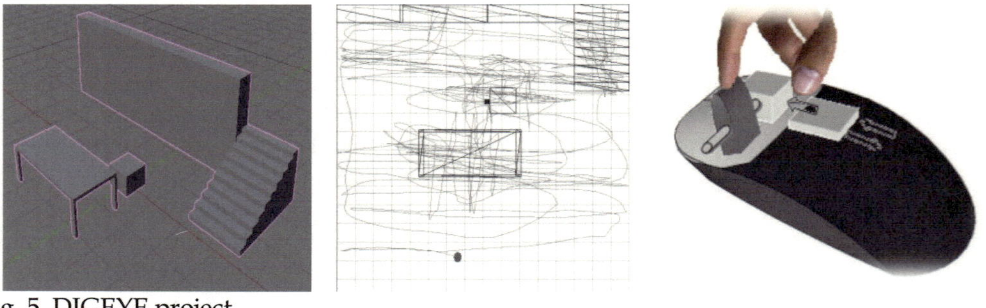

Fig. 5. DIGEYE project

The starting point of our work was purely dealing with tele-operation. We wanted to build a device to help navigation in environments where the visual information was not available. A prototype was achieved and tested. During these tests, we wanted to verify if people used with absence of visual information perform better than normal users. Following that, we decided to build a complete system allowing feeling the 3D geometry of the environment. We built this device including a stereo-vision system (for 3D modeling) and a haptic mouse with an actuated stick. The 3D geometry of a viewed scene is felt through the finger as a function of the mouse position and the stick height. We found that the information such device can display is not rich but enough for some tasks, namely, it allows to navigate safely. More recently and after brain activity signals analysis, we discovered that the device can be used as a diagnosis tool to diagnose some cognitive and brain pathologies. That is to say, our first problem, dealing with pure tele-operation issues leaded us to unexpected application.

6. Conclusions

In this document, I tried to give an overview of tele-operation: its history and origins, its background, its drawbacks and the perspectives it might offer to current researches and demands in robotics and other fields.

The construction of the future tele-operation/robotics is in progress and many people are still working on it. Surgery, prosthesis, rehabilitation, neurosciences, space, underwater and many other domains use this technology and the effort must be continuous.

The ideas described in this document are now facing the field reality through the experiments we are conducting. The preliminary results are encouraging, but a lot of work and a lot of efforts are necessary to progress.

I believe that tele-operation can help. However it cannot progress alone, it is necessary to work in multi-disciplinary way: tele-operation needs all the available knowledge, techniques and methods dealing with humans. We believe that technology alone cannot provide tools to achieve the robotics dream.

7. References

[1] Frédéric Kaplan. (2004). Les machines apprivoisées: Comprendre les robots de loisir Vuibert Ed, 2004.

[2] Rodney A. Brooks. (1990). Elephants Don't Play Chess Robotics and Autonomous Systems 6 (1990) 3-15

[3] Crevier, Daniel. (1993). AI: The Tumultuous Search for Artificial Intelligence, New York, Basic Books Edts, (1993).

[4]Vukobratovic, M. (1973). How to control artificial anthropomorphic systems. *IEEE Trans. Sys. Man Cybernet.* 3(5):497–507 (1973)

[5]Batya Friedman Peter H. Kahn, Jr. Jennifer Hagman. Hardware Companions? - What Online AIBO Discussion Forums Reveal about the Human-Robotic Relationship

[6]Erhan Oztop david w. Franklin, Thierry Chaminade. (2004). Human-humanoid interaction: is a humanoid robot perceived as a human? humanoids 2004

[7] Christoph Bartneck, Takayuki Kanda, Omar Mubin, Abdullah Al Mahmud. (2009). Does the Design of a Robot Influence Its Animacy and Perceived Intelligence? Int J Soc Robot (2009) 1: 195–204

[8] W. R. Ferrell. (1965). Remote Manipulation with Transmission Delay. *IEEE Transactions on Human Factors in Electronics*, 6:24–32, Sept 1965.

[9] Peter F. Hokayem Mark W. Spong. (2004). Bilateral Teleoperation: An Historical Survey Automatica 2004

[10] A. Bejczy, S. Venema, and W. Kim. (1990). "Role of computer graphics in space telerobotics: Preview and predictive displays," in *Cooperative Intelligent Robotics in Space, pages 365–377*, November 1990.

[11] T. B. Sheridan. (1993). "Space teleoperation through time delay: Review and prognosis," *IEEE Transactions on Robotics and Automation*, vol. 9, no. 5, October 1993.

[12] S. Zhai, P. Milgram. (1991). A telerobotic virtual control system Proc. SPIE, vol. 1612, Cooperative Intelligent Robotics in Space ii, Boston, nov. 11-13, 1991.

[13] Sören Krach, Frank Hegel, Britta Wrede, Gerhard Sagerer, Ferdinand Binkofski, Tilo Kircher. (2008). Can Machines Think? Interaction and Perspective Taking with Robots Investigated via fMR. PlosOne-July 2008, (open-access at http://www.plosone.org/article/info:doi/10.1371/journal.pone.0002597)

[14] Ackermann, E. (1996). Perspective-Taking and object Construction. In *Constuctionism in Practice: Designing, Thinking, and Learning in a Digital World* (Kafai, Y.,and Resnick, M., Eds.). Mahwah, New Jersey: Lawrence Erlbaum Associates. Part 1, Chap.2. pp. 25-37(1996).

[15] D. Weatherford. (1985). Representing and Manipulating Spatial Information from Different Environments: Models to Neighborhood. In: Cohen, R. (Ed.), The Development of Spatial Cognition, Lawrence Erlbaum, Hillsdale, NJ. pp. 41-70 , 1985.

[16] R. Cohen and Cohen. (1985). The role of activity in spatial cognition. In: Cohen, R. (Ed.), The Development of Spatial Cognition, Lawrence Erlbaum, Hillsdale, NJ. pp. 199-223. , 1985.

[17] Abderrahmane Kheddar. (2001). Teleoperation based on the hidden robot concept. IEEE Transactions on Systems, Man, and Cybernetics, Part A 31(1): 1-13 (2001)

[18] P. Willemsen, M. B. Colton, S. H. Creem-Regehr, and W. B. Thompson, (2009). The effects of head-mounted display mechanical properties and field of view on distance judgments in virtual environments, ACM Trans. Appl. Percept., vol. 6, no. 2, pp.1–14, 2009.

[19] Berthoz, A., Jorland, G. (2005). L'empathie, Eds Odile Jacob, 2005

[20] S. Tachi, H. Arai and T. Maeda. (1989). Tele-existence Visual Display for Remote Manipulation with a Realitime Sensation of Presence, Proceedings of the 20th International Syposium on Industrial Robots, pp.427-434, Tokyo, Japan (1989)

[21] P.Coiffet G.Burdea. (2003). Virtual Reality Technologies Wiley Eds (2nd edition), 2003

[22] Ee Sian Neo, Kazuhito Yokoi, Shuuji Kajita, Fumio Kanehiro, Kazuo Tanie: Whole body teleoperation of a humanoid robot -a method of integrating operator's intention and robot's autonomy. ICRA 2003: 1613-1619

[23] Hans Moravec. (1977). *Towards Automatic Visual Obstacle Avoidance* Proceedings of the 5th International Joint Conference on Artificial Intelligence, MIT, Cambridge, Mass., August 1977, p. 584

[24] S. Herbert. (1969). The science of the artificial, MIT Press, (1969)

[25] Metta G., Sandini G., Natale L., Craighero L., Fadiga L. (2006). Understanding mirror neurons: a bio-robotic approach Interaction Studies, 7(2), 197-232. 2006

[26] Pfeifer, R., Scheier, C., (2001). Understanding Intelligence (MIT Press, 2001)

[27] Ekeberg, Ö. (1993). A combined neuronal and mechanical model of fish swimming. *Biological Cybernetics, 69*, 363-374. 1993.

[28] H. Mayer, F. Gomez, D. Wierstra, I. Nagy, A. Knoll, and J. Schmidhuber. (2008). A System for Robotic Heart Surgery that Learns to Tie Knots Using Recurrent Neural Networks. Advanced Robotics, 22/13-14, p. 1521-1537, 2008

[29] R. J. Anderson and M. W. Spong. (1992). Asymptotic Stability for Force Reflecting Teleoperators with Time Delay The International Journal of Robotics Research.1992; 11: 135-149

[30] Ackermann, E. (1996). Perspective-Taking and object Construction. In Constuctionism in Practice: Designing, Thinking, and Learning in a Digital World (Kafai, Y., and Resnick, M., Eds.). Mahwah, New Jersey: Lawrence Erlbaum Associates. Part 1, Chap. 2. pp. 25-37, (1996).
[31] Piaget, J. and Inhelder, B. (1967). The coordination of perspectives In The child's conception of space, Chap. 8. pp.209-246. New York: Norton & Co. (1967)
[32] Fadiga, L.; Fogassi, L.; Pavesi, G.; Rizzolatti, G. (1995). Motor facilitation during action observation: a magnetic stimulation study Journal of Neurophysiology 73 (6): 2608–2611(1995),

Consideration of skill improvement on remote control by wireless mobile robot

Koichi Hidaka*, Kazumasa Saida† and Satoshi Suzuki‡
*Department of Electrical and Electronic Engineering, Tokyo Denki University, Japan
†Epson Co. Ltd, Japan
‡Department of Robotics and Mechatromics, Tokyo Denki University, Japan

Abstract

This paper considers the quantification of skill progress in order to measure a remote operational skill online. This method is very important to make a support system adapting to operator. This support system is called as Human Adaptive Mechatronics (HAM) System, and HAM was done as Center of excellent(COE) project promoted by Japan Society for the Promotion of Science.

Many approaches exist to attain this goal and human skill level depends on many factors such as condition, equipment and environment of operation. Therefore we first pay attention to delay time during the machine operation in order to aim at the acquisition of evaluation quantity on state of skill and we examine relationship between skill level and input/output delay time. For this analysis, we utilize a tele operated robot system to obtain data. We experiment with a simple task in which we operate a wireless mobile robot(WMR) with pinhole camera in order to reach the goal through the two check points. It is necessary that human operates machine by using image information from display monitor in the tele operation system. We analyze the correlation between operation date and WMR position based on these data. Next, we especially investigate a change of operational skill progress on curve path using data measured by simple task and analyze the data because curve control of tele operating robot only based on display is difficult to the operation in straight line path for human. For this purpose, we identified a delay of response with operation data based on ARX (AutoRegressive model with eXogenous input) model by using position and attitude data of the robot and then we analyze relationship and progress of skill level by using these data. A system operated by human can be considered as a closed loop system, and human can be regarded as controller in the system(3; 4). Furthermore, we classify operators into groups by correlation between delay time and total time of operation and area of stability poles and total time. And we analyze data by data of distance, total time and curvature and decide effective and important factors to skill level for each group. As these results, we consider skill parameter of tele-operations.

1. Introduction

We are living a comfortable life by using various gadgets of mechatronics products now. We use many machines in our work or life today and we usually operate the machines. There is not only simple machine system, such as cleaning machine, but also complex machine system, e.g. airplane system. Mechatronics is a key technology in our today's society. Mechatronics is known as the discipline integrated by mechanical, electrical and information technology and has been used to produce advanced artifacts used in modern society. The main issue of the mechatronics, however, is how to control the machine effectively. Mechatronics has been developing to design integrated systems that consider human and environment, recently. It is easy for me to manipulate the simple machine but it is difficult that we use a complex machine system safely. It is usually demanded that human learn an operation of the machine. Long term and much effort are needed to become skilled for human in many cases. The reason is that the operation system of the machine is not easy for human. Then an error motion causes unstable action of system even if the system with controller is stable(5). One of the reasons is that the machine does not adjust itself regardless of the human skill. That is, the ordinary machines were not designed to assist human to improve one's skill. Human can adapt to machine, sometimes make trouble in human–machine systems. To improve this wrong relation, mechatronics should pay attention to human skill level and adapt to the operator's skill ability assisting the operation.

Human Adaptive Mechatronics(HAM) is an intelligent mechanical system that adapts to human skill under various environments, improves human skill, and assists the operation to achieve best performance of the human machine system(1). In this new kind of human machine system, human factor has to be taken into consideration for the motion control design. The mechanical model of the skill-based human operation includes the various psychophysical limitations inherent in the human operator. Kleinman studied the dynamics characteristics of pilots(2). Generally, the beginner operator makes a mistake enough. On the other hand, the expert operates very well. For the reason, it is very important that the machine system can adapt the human skill. Our motivation of research is to find how we get good skill under this background. For the purpose, we make an experiment in this paper.

In considering a human as a control system, the system is treated as cascade of inherent part of reaction of time delay and lag attributed to the neuromuscular system where time delay comes from the various internal delay associated with visual or central process. Furthermore, it is aimed that the predictor models the human's compensation for his/she inherent time delay. This prediction is interpreted as Smith predictor(6). However, this delay time of controller for system can not be excluded from the response. This means that the delay time of operation can not be changed by training. Furthermore, mismatch of time delay causes the stability of operation (12) and the response is slow in the movement. Then we think that the delay time is important element of human operation. For this consideration, we first investigate the relation between the delay time and progress of operation using a WMR in simple experiment. Subjects operate the WMR with pinhole camera to reach a goal through two check points in the experiment. They manipulate by only using image of pinhole camera under the assumption that human regulates his/her delay time between input signal and output response. In this experiment, the input signals make position and angle of joystick and the output responses are measured by other camera system. The operation signal of robot is analog so that this signal gives linearity between joystick positions or angles and velocity signals of WMR. We can feel the linear response between joystick operation and WMR moving and we do not sense the time delay because of quick response. We analyze the position and velocity of WMR and the operation log of joystick. We calculate correlation with start input signal and the response of WMR to the input. From the correlation data, relation between time to goal and the distribution of delay time. Next, using the data based operation data on curve, we identify a delay of response by using ARX model. Since operation in this experiment is more difficult than in straight path, we consider that the operation data in curve include delay of response.

The subjects are classified into some groups by using data based on the relationship between maximum stability pole of ARX and operation time and between delay time and the operation time. Furthermore, detail analyses are done to the subject by whom each group has a strong correlation in each relation and we multilaterally examine the delay of response and the relation of the advancement of the operation.
This paper is organized as follows. In section II, tele operation control system of machine and response delay of human operation in the system is introduced. We explain an experimental system that consists of tele operated wheeled mobile robot and measurement system in section 3. In this section, we first introduce a test field of experiment and manipulation. Section 4 explains an analytical method of the experiment data. First of all, we examine each delay time of seven subjects based on operational data of WMR. We use of the data operated in a special environment in order to examine the relation between response delay and operation mistake. These data were measured in the situation which the wall is near. Next, the delay of the operation based on a curve running data is shown by using ARX model. The subjects are being classified into some groups and these data are analyzed by using Independent component analysis in this section. Last section 5 is conclusion and discussion.

2. Human model and skill in tele operation control system

Our motion control system is composed of the brain, the outside environment, and the body. The brain corresponds to the control center, and the body receives information with the outside and acts actually. Therefore the skill level of human operation has relation to an environment and equipment of operation strongly and this relationship is greatly important in the analysis of the motion control system. To achieve a desirable motion control corresponding to the environment, an internal model of external dynamics that consists of the body and the environment is needed(10). On the other hand, there are transmission delays of the nervous system in our sense and movement. This delay has the range from 300ms to 10ms and the delay in the feedback loop makes the system unstable. Therefore, to do a smooth and fast motion, the decrease of the influence of the response delay is important and the forecast of environmental change into action is needed (11). A mechanism to compensate to time delay is explained as so–called Smith Predictor. In the machine operation that should consider time delay, there is a tele operation system(6). In the case of tele operation system, real time camera image is significant for the judgment and the response of the visual feedback is late. Thus it is important to adjust the delay for good motion control. Fig. 1 shows a block diagram of a system to which human does a tele operation of machine. While this method is to cognitive steady process, these modeling methods tend to become subjective because the method needs analysts cognitive judgment. Based on this viewpoint, we consider two type skills such as cognitive and operation skills are needed to operate a machine system and we examined the relation between environment recognition and process of operational skill in tele operated system(8). From this result, cognitive skill have little relevancy in simple task of machine manipulation and operation skill is more important than cognitive skill for adult. In an operation task of a machine, on the other hand, the delay in operation affects the results.
In the system by which human operates a machine, the machine can be considered and the plant and human be regarded as a compensator. When delay time is in the closed system, the effect of control input appears late and a high gain makes the feedback system unstable(12). The time delay e^{-sL} does not change a gain of system but only change a phase of system and degrease in time delay L increases a stability margin of phase. From this result, the time delay shows the possibility of control performance in tele operation system.

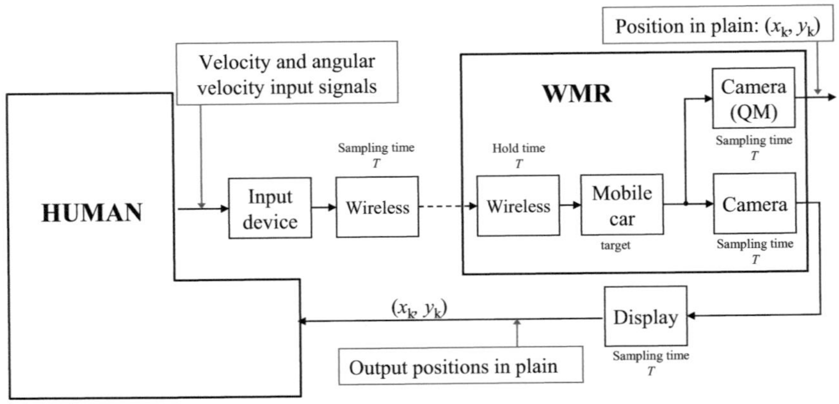

Fig. 1. Block diagram of human/machine system

3. Experiment for data acquisition for analysis

A. Test field

In order to analyze time delay based on the data between velocity command v and the actual translational velocity of WMR, the data is calculated from position data of the WMR measured by stereo vision tracking system; Quick Mag. The WMR used in this experiment is shown in Fig. 2(left). The driving device is used by stepping motor and a pinhole camera is mounted forward in order to give the operator the front image of the WMR. A joystick is used to operate this WMR. This joystick signal is analog so that signal gives linearity between joystick positions or angles, and velocity signals of the WMR. We can feel the smooth response from joystick operation to WMR motion. The test field is a small maze that consists of block–wall, start point, goal point, and several check points. Operator manipulates the WMR by seeing the image from the camera, and moves the WMR from start point to goal point passing the intermediate check points by using joystick. Right of Fig. 2 shows the test field. In this experiment, operator has to manipulate the WMR by only using image information of position and movement. We investigate the correlations such as the correlation coefficients between delay time of angular velocity, pole estimated ARX based on angular velocity data and total time by using the data from a start position of analysis to goal shown in Fig. 4. This area is narrow and subjects have to turn to go into the goal in this place. Therefore, many subjects can not operate a wheeled mobile robot (WMR) well in this place. From this reason, we consider that rotation operation in narrow place is more difficult than the straight advancement operation and analyze the angular velocity data of this area. We evaluate skill by total time of each trial. Before the experiment, subjects were permitted to see the test field, and were demanded by an experimental instruction so that they might operate the WMR to the final goal as fast as possible. Ten trials were imposed to each subject. Position and rotation of the WMR is measured by a real time visual motion capture system: QuickMag with sampling time 16[ms].

B. Operation signal of wheeled mobile robot

An operator manipulates a joystick by using front–back direction y and rotational angle z. Fig. 3 shows operation of the joystick. Translational velocity and rotational velocity are computed as

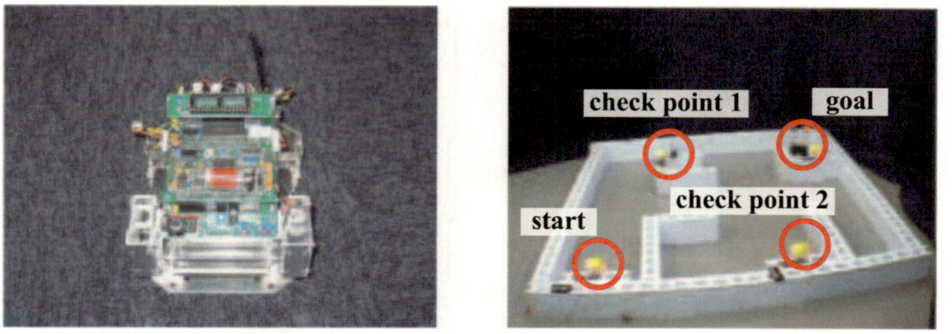

Fig. 2. Wheeled mobile robot(left) and maze test filed(right)

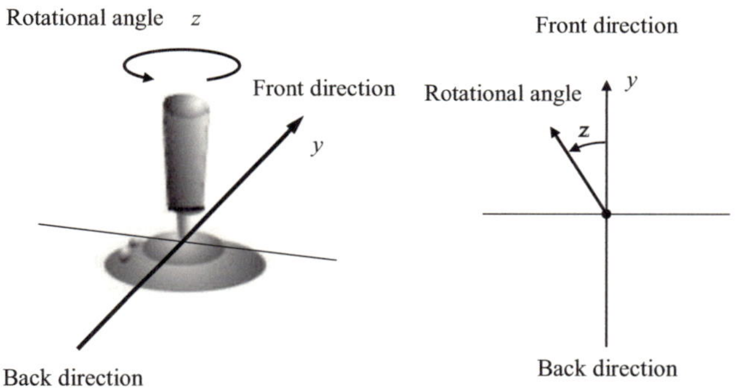

Fig. 3. Joystick movement

$$v(t) = \begin{cases} \text{sign}\,(y(t)) \left(\dfrac{v_{max} - v_{min}}{1000 - 400} \left(|y(t)| - 400 \cdot \text{sign}\,(y(t)) \right) + v_{min} \right) & \text{if } 400 \leq |y(t)| \leq 1000 \\ 0 & \text{if otherwise} \end{cases} \tag{1}$$

$$\omega(t) = \begin{cases} \text{sign}\,(z(t)) \left(\dfrac{\omega_{max}}{800} \left(|z(t)| - 200 \cdot \text{sign}\,(z(t)) \right) \right) & \text{if } 200 \leq |z(t)| \leq 1000 \\ 0 & \text{if otherwise} \end{cases} \tag{2}$$

where $v_{max}[m/s]$, $v_{min}[m/s]$ and $\omega_{max}[deg/s]$ are given by 0.5, 0.125 and $(v(t) - v_{min})/90$ respectively and these values are tuning parameters. Using these input signals, the velocity of right and left wheel such as $v_r(t)$ and $v_l(t)$ of WMR are calculated by

$$v_r(t) = \frac{2v(t) + 90\omega(t)}{2}, \; v_l(t) = \frac{2v(t) - 90\omega(t)}{2} \tag{3}$$

This joystick has free area in position and angle and the input signal gives as 0 in the area. The operator of robot is analog so that this signal gives linearity between joystick positions or angles and velocity signals of WMR. Then operator can feel the linear response between joystick operation and WMR moving and he does not sense the time delay because of quick

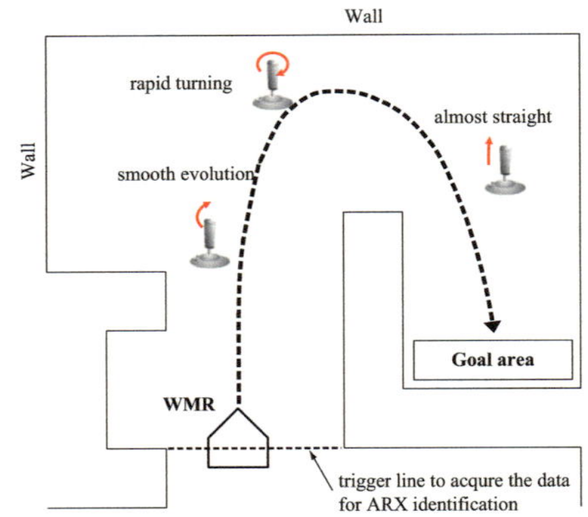

Fig. 4. Operation track of WMR from start position of analysis to goal area

response. The output responses are measured by QuickMag in this experiment. The subject tried this operator 10th times. We define the progress of skill as arrival time. Because of the definition, the subject were demanded to reach the goal as soon as possible before this experiment. We calculate the correlation between input signal and output response by using these experiment data. In this experiment, we assume that operators tend to predict the motion of WMR through the image information and he/she adjusts the delay time of output response. Under the assumption, we think that changing of delay time can use the estimation of skill level and investigate the tendency of decrease of goal time with concentration of time delay. The operation time is used to judge the skill level in this experiment. Then we understand that the faster the goal WMR can be arrived at, the more he/she became good at his/her operation.

4. consider of relationship between skill level and response delay time

4.1 Analysis based on data between checkpoints

We first searched the delay times between input signals and output responses based on the data for position and direction of the center of gravity. A male participated in this experiment. In this experiment, we used the positive changing date and counted the number of delay time(7). The data varying from 0 to nonzero number and relating the response of WMR to the input signal can be found. Using this idea, we calculated correlation data of input u(t) and output y(t) such as

$$\lim_{t \to \infty} \frac{1}{T} \int_{-T/2}^{T/2} u\,(t)\,y\,(t+\tau)d\tau \tag{4}$$

The input data such as joystick position, angle and direction of WMR are given by digital data. Then we calculate the correlation as the number of $\sum_{l=-0}^{T} u(k)y(k-l) \neq 0$. Fig. 5 gives a result of reaching time. (1) is the total time from start to 1st check point, (2) and (3) experiment the total time from 1st check point to 2nd check point and 2nd check point to goal. (4) is time to total operation time. These results indicate that operation decreases with increase the step of operation. We consider that operator tends to manipulate the handle of joystick well and that skill level in 10th experiment is higher than in 1st experiment.

Fig. 6 to 13 show the results for number and rate of delay time of velocity. Fig. 6 and Fig. 7 give the results of delay time of velocity and rate of delay time in 1st operation. The data in the

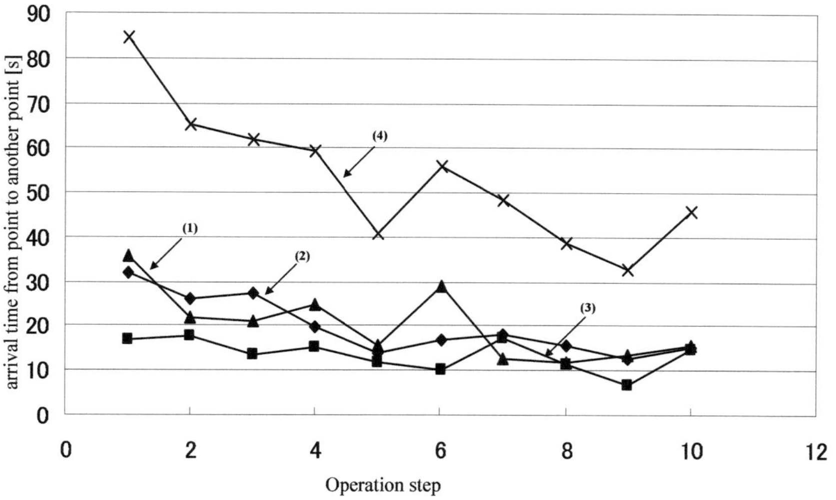

Fig. 5. Arrival time from point to another point;(1) start to point 1, (2) point 1 to point 2,(3) point 3 to goal

first operation show that delay time distributes wide range such as about 150ms to 800ms and there are two peaks at near 300ms and 500ms. On the other hand, we can find out the features in Fig. 8 and Fig. 9 that range of delay time is closer than first result and the time concentrates two times such as 300ms and 500ms. By comparison with 1st operation, improvement of operation makes the distribution small.

Fig. 10 to 13 present the result of delay time in angular velocity. These times are faster than time for velocity because we move joystick in operation of angular velocity. These results also give that the range of delay time changes. In first operator, the delay time appeared over two times exist in near 200ms to 350ms and the delay time over same times concentrate near 350ms in the 10th operation. We consider from these date that human does not accustom him to operation with image of camera and he can not estimate next move of WMR. Then he hardly operate joystick at good timing and he has many delay time. Otherwise he learns good timing through many operations and the delay in 10th experiment does not expand wildly. We think the distribution of delay time relates with skill level.

4.2 Analysis on progress process of operational skill level by ARX model

Next, we analyze a progress process of operational skill level by ARX model given by equation(5). Operated robot and operating environment used the same one as experiment of section 4.1. Moreover, the inadequate data set was rejected from the identification process to keep reliability of the identification, because inadequate data set, e.g. collision against a wall, does not include correct movement of WMR to operational command.

In this example, seven adult male participated in this examination, whose age range are 22 to 24. subjects are demanded to reach a goal as fast as possible. For the reason, we consider that information of the total time can be used as index of skill–level. Results of total time of the subjects are shown in Fig. 14. These results indicate that total time of subjects monotonically degreases as trial increase. we can consider all subjects have improved the operation level from these results. The data estimated delay time by ARX is the same as the data used in

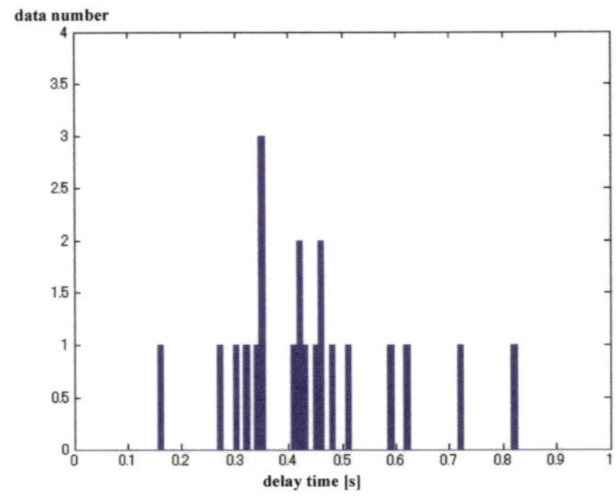

Fig. 6. Result of delay time for velocity in 1st operation

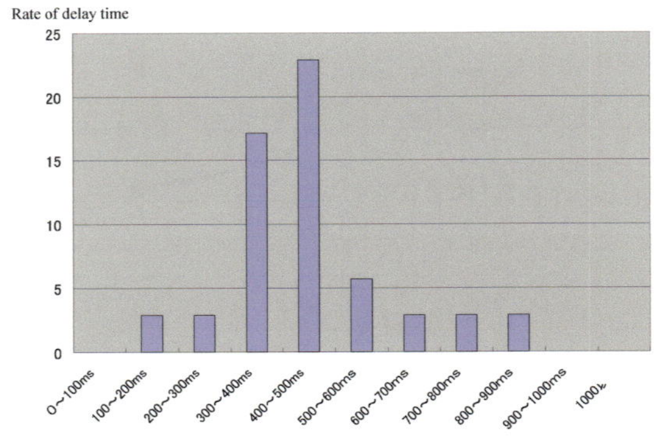

Fig. 7. Rate of delay time for velocity in 1st operation

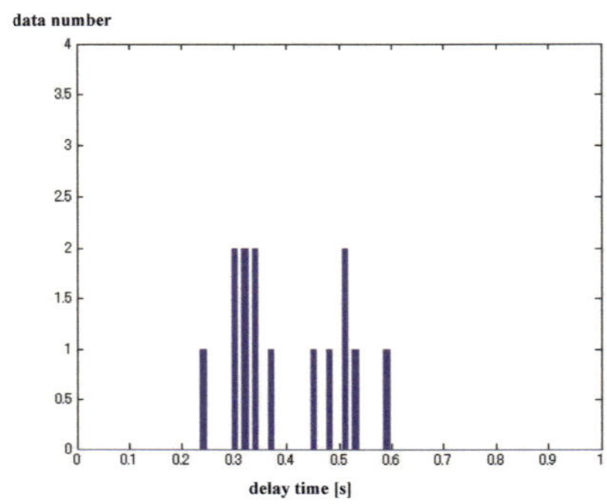

Fig. 8. Result of delay time for velocity in 10th operation

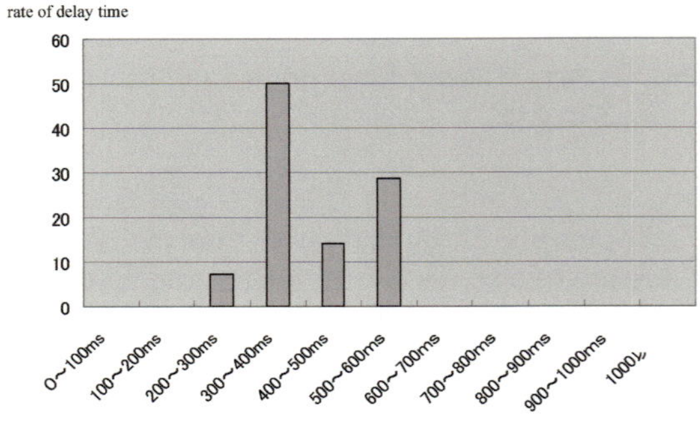

Fig. 9. Rate of delay time for velocity in 10th operation

section 4.1. That is, the input signals are given by (1) and (2) and the output is actual translation velocity of a WMR that is computed from position data measured by a QuickMag. Using data given in experiment, time delay is estimated by searching minimum identified error by equation (5) with identified parameters. The orders, n and m are specified as both 5 because the suitable value can be decided by using loss function given as $\sum_i^{N/2} \epsilon^2 (i, \theta)$ where ϵ is computed by prediction error and N and θ are a number of data and data vector, respectively(9).

4.3 Validity of ARX model

Human characteristic of tele operation system is estimated by using ARX model with delay in this paper. ARX model is given by

$$y(t) + a_1 y(t-1) + \cdots + a_n y(t-n) = b_0 u(t-L) + b_1 u(t-L-1) + \cdots + b_m u(t-L-m) \tag{5}$$

where n, m are degree of an ARX–model, and L is time delay. Similarly, for rotation model, the input is angular velocity of a WMR. The characteristic of seven subjects were analyzed by

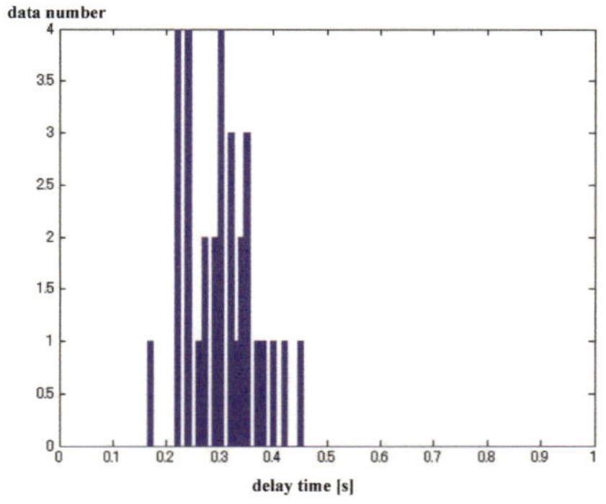

Fig. 10. Result of delay time for angular velocity in 1st operation

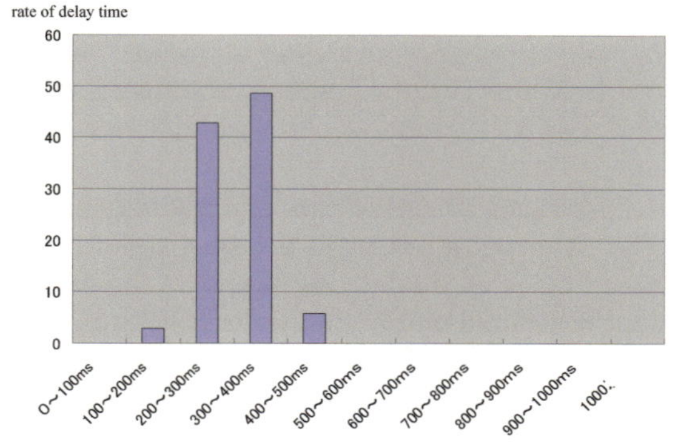

Fig. 11. Rate of delay time for angular velocity in 1st operation

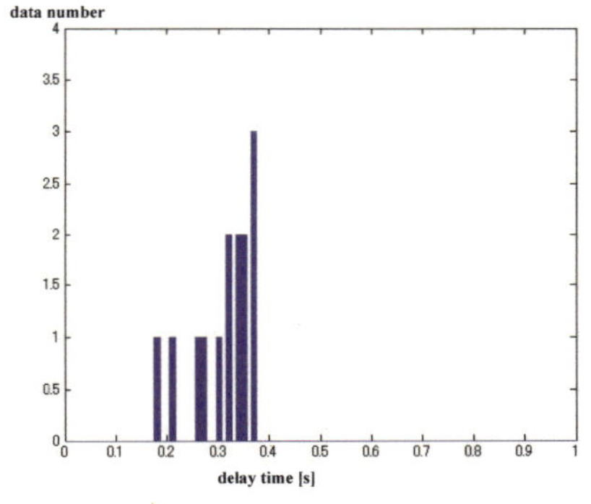

Fig. 12. Result of delay time for angular velocity in 10th operation

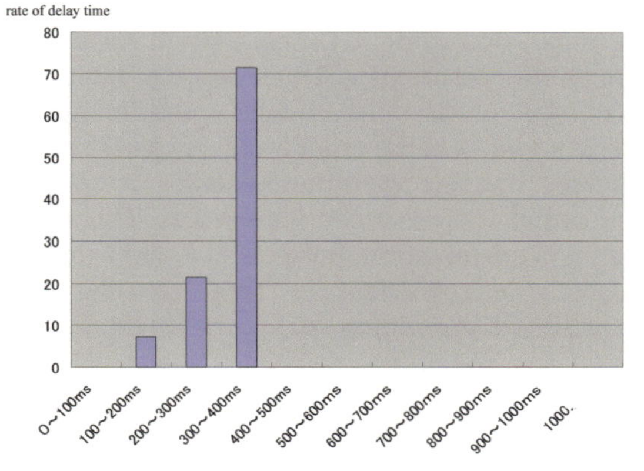

Fig. 13. Rate of delay time for angular velocity in 10th operation

using ARX model based on experiment data. This experiment field is shown in Fig. 4. First of all, we examine validity of ARX as human operation model. We compare output data with estimated data made by ARX. Although straight course can confirm forward scenery from a front camera, a forward situation in the curve course changes one after another through a front camera. For the reason, an operation of WMR in the curve course is more difficult. Since rotational operation is difficult and skill progress of each operator is different, we investigated not the translational velocity data but the rotational velocity data. The operational signal is calculated based on another data by using this ARX model after a coefficient of ARX model was estimated by using the data of curve path. That is, the data from a start position to goal shown in Fig. 4 are used for estimation of ARX model. The validity of this model is examined by the difference between an output of ARX and an actual output. Fig. 15 shows angular velocity that measured by QuickMag and estimated by ARX model, where x-axis and y-axis are an operation time and angular velocity. The large difference is not seen from Fig. 15 in the output, and the model also reproduces operation change well. Furthermore Human are usually stability controller and the ARX which is a model of human operation had to be steady.

Fig. 16 shows zeros and poles of the ARX model. This result of Fig. 16 shows that poles are in stable area and the model is stable. On the other hand, zeros are on the circumference of one in radius and the position of zeros affects an operational performance. As a result, we can find that ARX model is reasonable as an estimated model.

4.4 Analysis on delay time and stability pole of ARX

The data measured from the start point of the curve to goal point is used for the analysis based on ARX model. Fig. 4 shows the measure point. There is a wall and subjects have to start rotational motion in this area. This area is narrow and subjects have to turn to go into the goal in this place. Therefore, many subjects can not operate WMR well in this place. From this reason, we consider that rotation operation in narrow place is more difficult than the straight advancement operation. We think that the skill of operation appears more clearly in difficult operation areas, and we use the angular velocity data observed in this place. The Values as the correlation coefficients between delay time of angular velocity, poles of estimated ARX based on angular velocity data and total timeĄ@are used for the data analysis. Using these values, we try to classify the subjects into two groups. Table 1 and table 2 show the two groups. According to the correlation coefficients l_ω and σ_ω, A and D have positive correlation for both coefficients. These subjects have a tendency such that operation mistake such as collision to wall decrease with operation number and these subjects tend to be able to operate WMR well. On the other hand, subject C has negative correlation coefficients σ_ω and l_ω. The subject has few operation mistakes in the first operation. However, the subject makes a lot of mistakes since second operation. This subject tends to reach goal as fast as possible since second examination. Subjects E, F and G have positive correlation coefficient σ_ω but negative correlation coefficient l_ω. They have the same tendency as the subject A and D. Then the coefficient correlation, p_a, p_g, between delay time and poles to subject A and G, who have

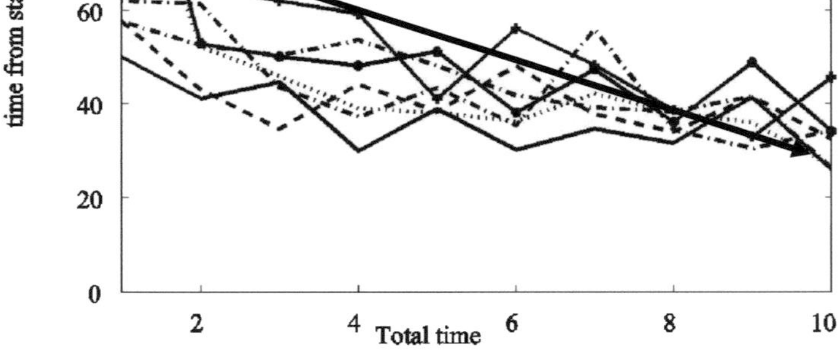

Fig. 14. Total time on each trial

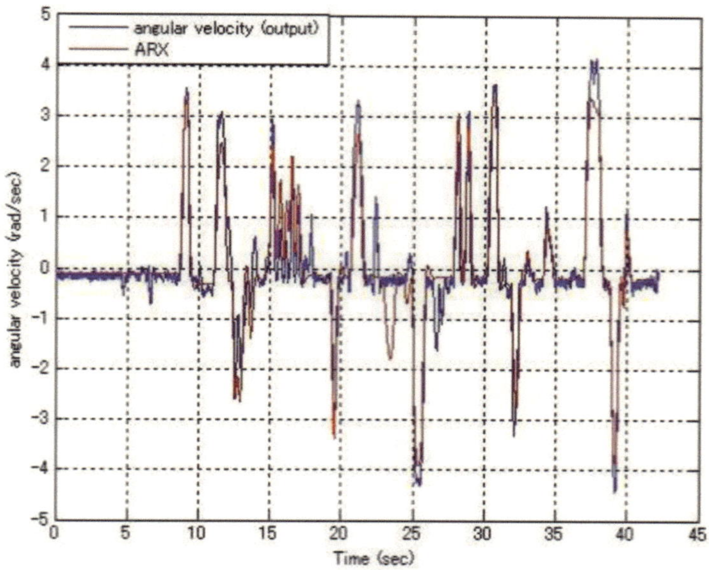

Fig. 15. Measured angular velocity of WMR and simulated response with estimated ARX model output

a maximum and a minimum correlation l_ω in the group (A,D) and (E,F,G). The correlation coefficients to A and G are computed as $p_a = 0.35$ and $p_g = 0.50$, respectively. p_g is larger than l_ω of G. Fig. 18(lower) shows the tracking path of WMR of subject G. Subject G tends to turn WMR smoothly without stopping around corner of maze. As the change to smooth path, the skill level of subject G progresses clearly. The progress of operation can also judge from the result of collision number shown in Fig. 20. The delay times of subject G shows in Fig. 19(lower). Therefore, subject G tends to decrease delay time with stability and the operation is steady as a result. On the other hand, for subject A, l_ω is larger than p_a. The delay times of subject A shows in Fig. 19. The result shows that the delay time is larger in first trial. The correlation coefficient l_ω of subject A by using data from second to tenth trial is computed as -0.036. In the operation since the second times, it can be said that the operation is steady. The feature can be found from Fig. 18. However, Results of Fig. 19 show that delay time and pole of angular velocity does not decrease monotonically. We consider that subject tends to do a same operate, so that the coefficients do not decrease. It can be confirmed that the operation track of WMR has the same tendency as straight line and rotation shown in Fig. 18(upper). On the other hand, subject C has negative correlation coefficients σ_ω and l_ω.

Next, the subjects A,B,C and D are also examined from the views of the values as (i) total time and variation of delay time, (ii) total time and kurtosis of delay time and (iii) total time and skewness of delay time. The subject A and D show the feature of $l_\omega > 0$ and $\rho_\omega < 0$, and the B and C have $l_\omega > 0$, $\rho_\omega < 0$ and $l_\omega > 0$, $\rho_\omega < 0$, respectively. Table 3 and table 4 show these values. For the values (i), (ii) and (iii), the other subjects except B do not have all positive values. However The definite feature of operation can not be found from these values. Then we investigate data from another viewpoint again.

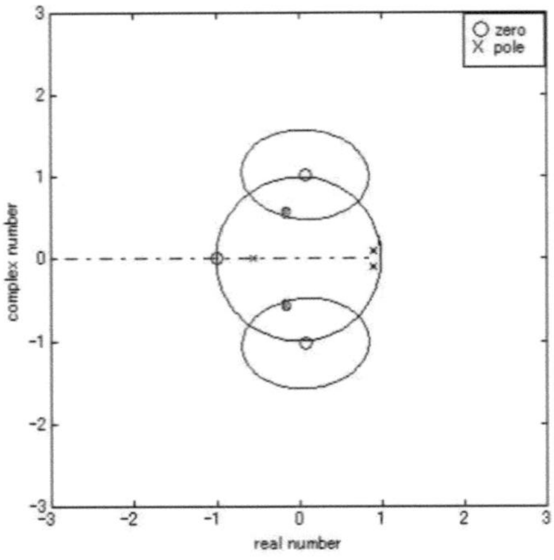

Fig. 16. Zero (○) and pole (×) of estimated ARX model

We classify the seven subjects by relationship between delay time and total time and by between maximum stability pole and total time. Fig. 27 indicates the relationship of the correlations. x axis is correlation coefficient of delay time and total time, and y axis is a greatest stability pole of ARX and total time. The subjects experimented ten times and sum of operation time was recorded. Results in Fig. 27 indicate that subjects can be divided into two groups. A group 1 includes C and D and B, F, and G are elements of a group 2. A group 1 shows the subjects have positive correlation. On the other hand, group 2 tends to a negative correlation. Subject A shows a tendency of increase, but regarded as the other group so that a rate of increase is different from group 1. In addition, subject E is included in both group 1 and group 2, but we include E into group 1 this time. We analyze the following terms two groups by using input and output data from 2nd point to goal, i.e., (i) distance from 2nd point to goal, (ii) operation time, (iii) the number of time that curvature has more than 5 and (iv) a rate of operation time from 2nd point to goal to total operation time. Fig. 28 and Fig. 29 are results on (i) to (iv) for subjects in group 1 and 2. Group 1 tend to decrease (i), (ii) and (iii) according to increase of experimental number of times. On the other hand, the curvature of subjects F and G in group 2 do not vary and their distance do not tend to decrease. For these results, the subjects in Group 1 which have positive correlation come to progress their manipulation skills because their stability and delay time of response are decreased with experiments times.

$l_\omega > 0$	subject A ($l_\omega = 0.57$)
	subject D ($l_\omega = 0.17$)
	subject B ($l_\omega = 0.02$)
$l_\omega < 0$	subject C ($l_\omega = -0.6$)
	subject G ($l_\omega = -0.14$)
	subject F ($l_\omega = -0.11$)
	subject E ($l_\omega = -0.003$)

Table 1. Classification table by delay time of angular velocity

$\sigma_\omega > 0$	subject G ($\sigma_\omega = 0.42$)
	subject F ($\sigma_\omega = 0.35$)
	subject A ($\sigma_\omega = 0.34$)
	subject D ($\sigma_\omega = 0.24$)
	subject E ($\sigma_\omega = 0.021$)
$\sigma_\omega < 0$	subject C ($\sigma_\omega = -0.51$)
	subject B ($\sigma_\omega = -0.15$)

Table 2. Classification table by pole of angular velocity

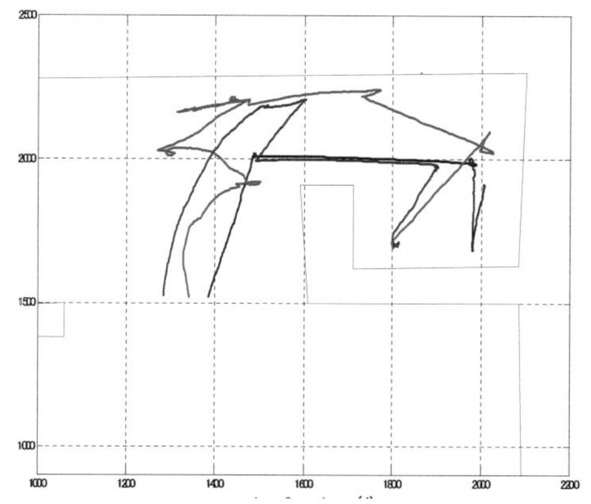

Fig. 17. Operation trajectory of subject A; 1th (blue), 5th (green) and 10th (red) trial

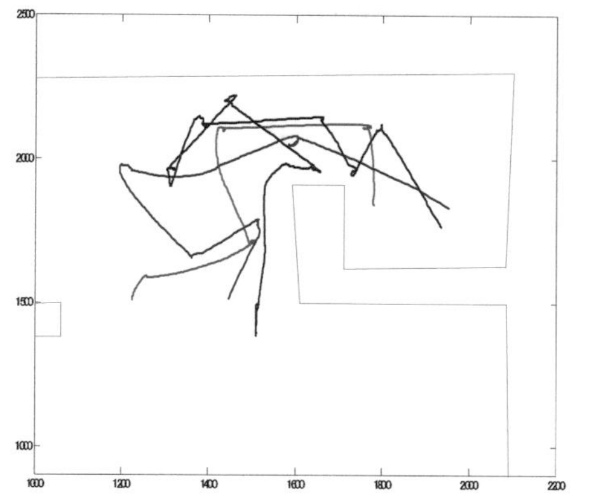

Fig. 18. Operation trajectory of subject G; 1th (blue), 5th (green) and 10th (red) trial

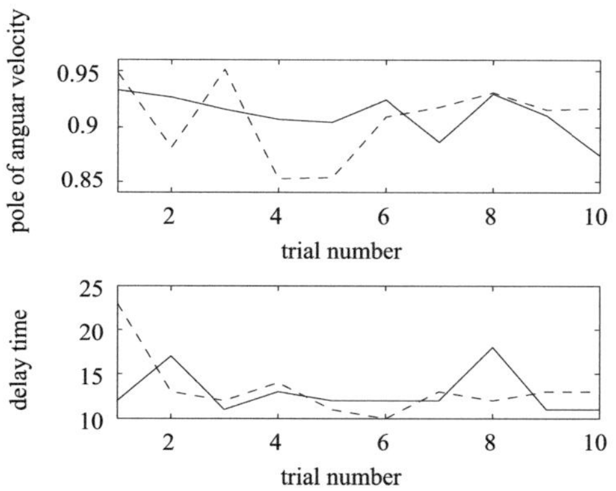

Fig. 19. Pole of angular velocity(upper) and delay time (lower) subject A (dashed line) and G (solid line)

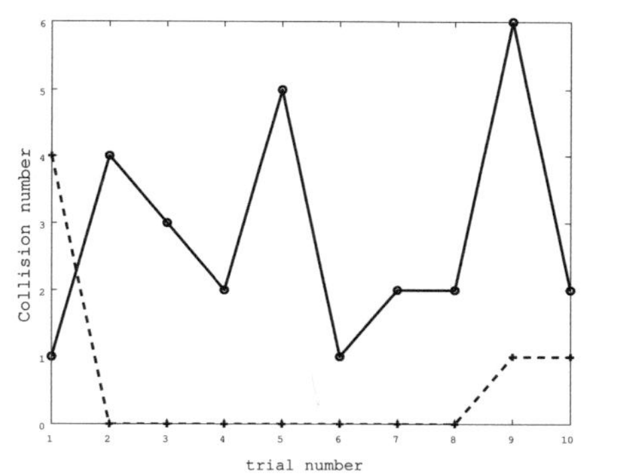

Fig. 20. Collision number of subject: A (solid line) and subject G (dashed line)

We consider that they can arrive at the goal only by a small number of turn operations. The subjects in Group 2 with negative correlation have the feature that stability margin tend to become more large with operation times, the response time, however, does not decrease. So we consider they tend to do useless operation, so that the distance and delay time does not decrease. Next, we examine each correlation value of these elements; i.e. (i) and (ii), (i) and (ii), (ii) and (iii). Each value is included in table 5. The each correlation of (i) and (ii) and (ii) and (iii) shows high values. We notice this result, and we demonstrate each operation path of C,D and G by using data from 2nd point to goal. Fig. 31 to Fig. 32 are results of operation path. When the operation times increases, each operation of subject D and C tends to be steady. On the other hand, The subject G of group 2 does not operate WMR stably every time. Fig. 18 is the result of operation of G. Moreover progress of operation differs in the subject C and D. D can rotate WMR at curve well and his path is smooth, but C does many useless manipulations. Furthermore as shown in table 1 and 2, it is important point that both l_ω and ρ_ω of C are positive.

	A	B	C	D
(i)	-0.330	-0.588	0.200	0.102
(ii)	0.138	0.506	-0.049	0.396
(iii)	-0.238	0.510	0.303	-0.190

Table 3. Correlation between total time and each data of velocity

	A	B	C	D
(i)	-0.160	0.006	0.020	0.090
(ii)	-0.067	0.335	-0.160	0.067
(iii)	-0.450	0.267	0.536	0.031

Table 4. Correlation between total time and each data of angular velocity

(i) total time and variation; (ii) total time and kurtosis; (iii) total time and skewness

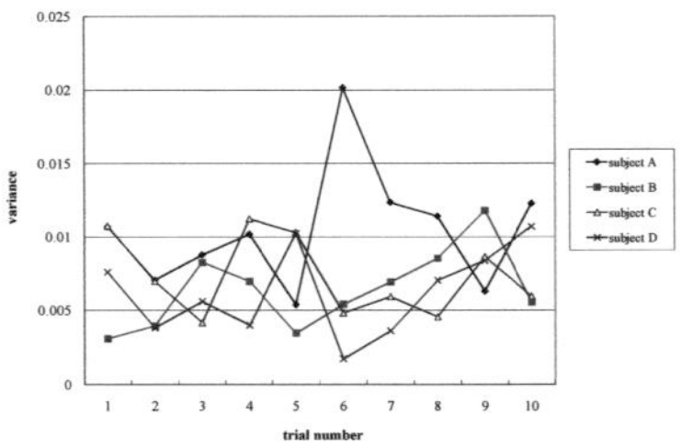

Fig. 21. Variation of delay time of velocity

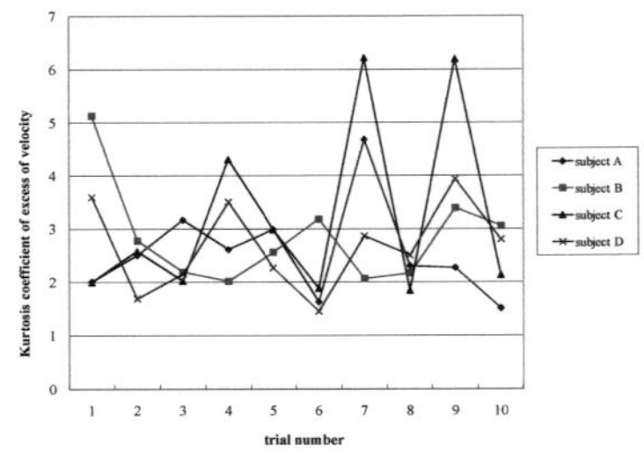

Fig. 22. Kurtosis of delay time of velocity

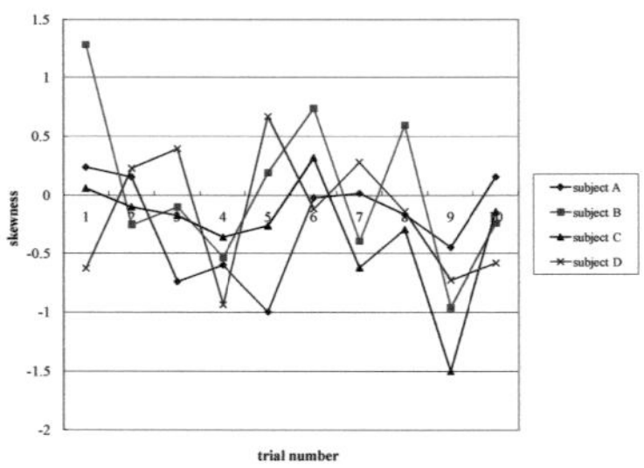

Fig. 23. Skewness of delay time of velocity

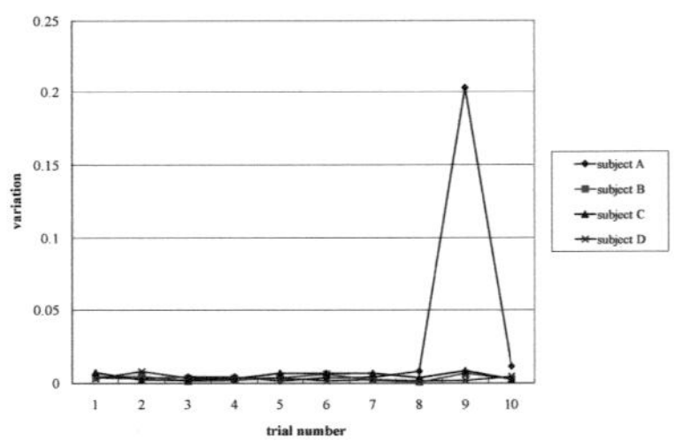

Fig. 24. Variation of delay time of angular velocity

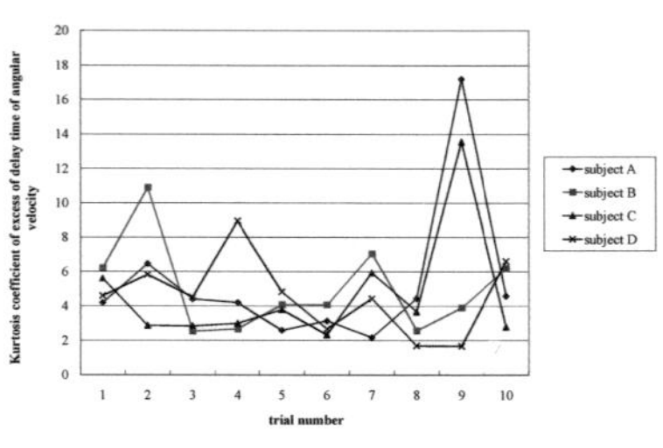

Fig. 25. Kurtosis of delay time of angular velocity

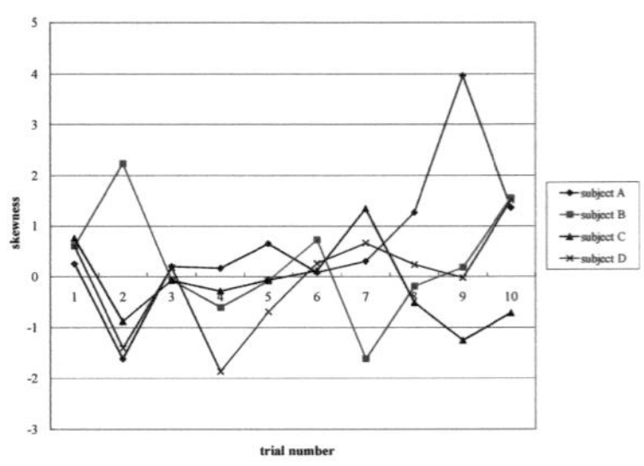

Fig. 26. Skewness of delay time of angular velocity

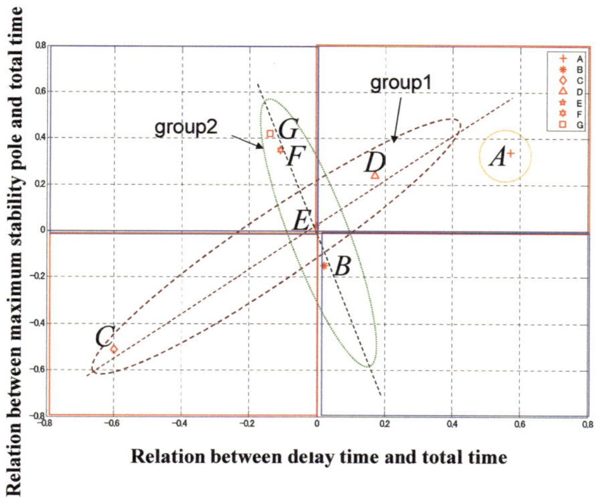

Fig. 27. Relation between delay time, maximum stability pole and total time

Table 5 Correlation of trajectory distance , path time and curvature on curve path

	Distance and time	Curvature and distance	Curvature and time	
C	0.81	0.76	0.93	g roup 1
E	0.93	0.8	0.92	
D	0.9	0.65	0.67	
B	0.83	0.77	0.86	g roup 2
F	0.8	0.87	0.85	
G	0.95	0.94	0.96	
A	0.64	0.49	0.53	

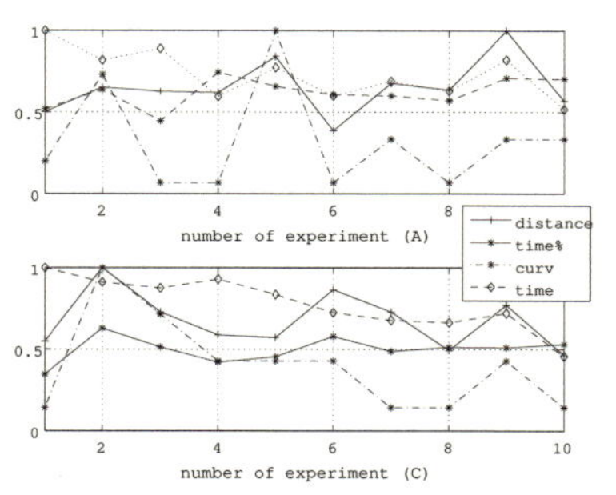

Fig. 28. Results of (i) – (iv) with subjects A, C;(i) distance (solid line), (ii) normalized operation time solid line with $*$), (iii) curvature (dot–dash line), (iv) normalized total time dashed line

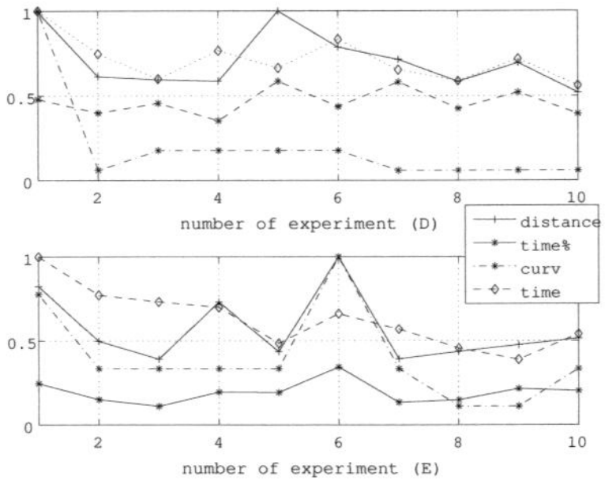

Fig. 29. Results of (i) – (iv) with subjects D, E;(i) distance (solid line), (ii) normalized operation time solid line with $*$), (iii) curvature(dot–dash line), (iv) normalized total time dashed line

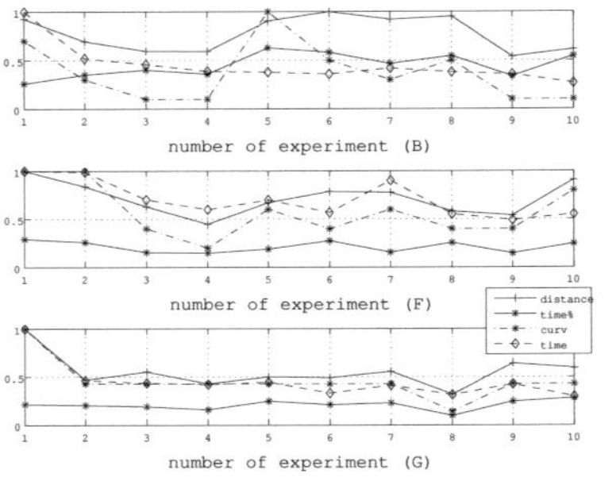

Fig. 30. Results of (i) – (iv) with subjects B, F, G;(i) distance (solid line), (ii) normalized operation time solid line with $*$), (iii) curvature(dot–dash line), (iv) normalized total time dashed line)

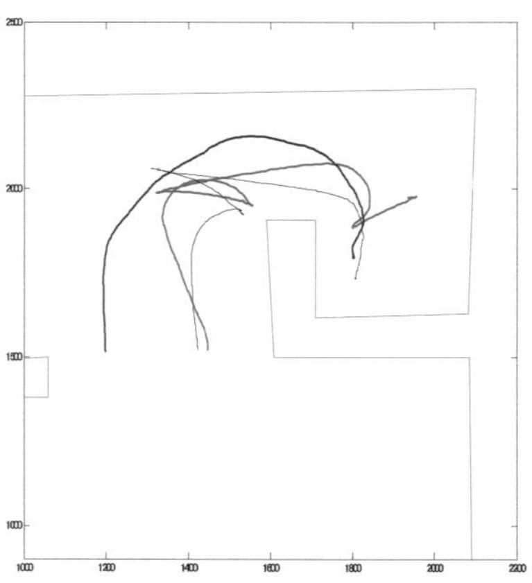

Fig. 31. Operation trajectory of subject D; 1th(blue), 5th(green) and 10th(red) trial

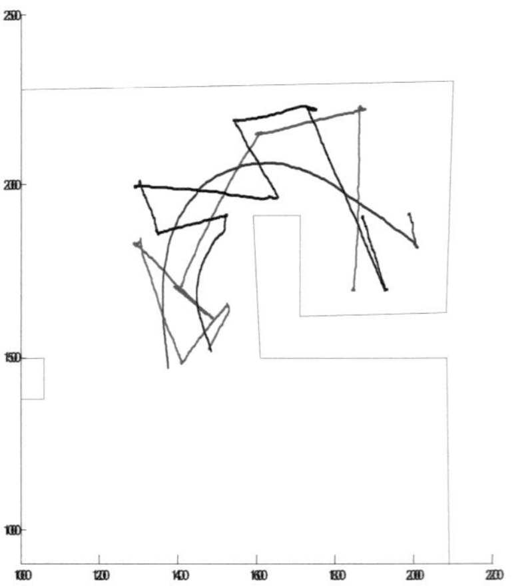

Fig. 32. Operation trajectory of subject C; 1th(blue), 5th(green) and 10th(red) trial

5. Conclusion

We analyzed correlations such as the correlation coefficients between delay time of angular velocity, pole estimated ARX based on angular velocity data and total time. We can classify the subjects into two groups for each coefficient based on the correlation coefficients l_ω and σ_ω, where σ_ω and l_ω signify the coefficient between pole of ARX, delay time of angular velocity and total time, respectively. One group has positive correlation for both coefficients. On the other hand, One in the other group has negative correlation coefficients σ_ω and l_ω and the other subjects in the second group have positive correlation coefficient σ_ω but negative correlation coefficient l_ω. They have the same tendency as the some subjects in first group. Next, we analyzed the coefficient correlation, p_a, p_g, between delay time and poles of subjects who have same tendency in both group. We found the following tendency based on the correlation coefficients to the subjects in two groups;

(1) Decrease of delay time tends to the stability of operation.

(2) Operation track has the same tendency when l_ω is larger than p_a.

Furthermore, we classified 2 groups by correlation of coefficient with operation time and we considered the skill levels of the groups based on rotation manipulation and time from 2nd pint to goal. The results indicates that the group that tends decrease of response delay decreased a path distance and manipulation time also. For these results, the response delay is one of feature for skill level and the quantity is useful for inference of skill level.

Acknowledgments

This research was supported by the Grant-in-Aid for 21st Century COE (Center of Excellence) Program in Ministry of Education, Culture, Sports, Science and Technology in Japan. The authors would like to thank all COE members.

6. References

[1] S. Suzuki, F. Harashima and Y. Pan, "Assistant control design of HAM including human voluntary motion," in Proc. 2nd COE Workshop on HAM, TDU, Japan, pp.105–110, March 2005.

[2] D.L. Kleinman, S. Baron and W.H. Levison, *"An Optimal Control Model on Human Response Part I:Theory and Validation,"* Automatica, vol.8, no.6, pp.357–369, 1970.

[3] Mitsuo Kawato, "Internal models for motor control and trajectory planning," *Motor systems*, 1999, pp.718–727.

[4] M. Kawato, K. Furukawa and R. Suzuki, "A hierarchical newral–network model for control and learning of voluntary movement," Biological Cybernetics. vol.57, pp.169–185, 1987.

[5] M. Pachter and R. B. Miller, "Manual Flight Control with Saturating Actuators," *IEEE Control Systems*,pp. February, pp10–20, 1998.

[6] R.C. Miall, D.J. Weir, D.M. Wolpert and J.F. Stein, *"Is the Cerebellum a Smith Predictor ?,"* Journal of Motor Behavior, vol.25, no.3, pp.203–216, 1993.

[7] Kazumasa Saida, Eizi Kodama, Yorito Maeda, Koichi Hidaka, Satoshi Suzuki, *"Skill Analysis in Human Tele-operation Using Dynamic Image,"* IEEEIndustrial Electronics,IECON 2006, pp.4528–4533, November, 6–10, 2006.

[8] Yorito Maeda, Satoshi Suzuki, Hiroshi Igarashi, Koichi Hidaka , " Evaluation of Human Skill in Teleoperation System," SICE–ICASE International Joint Conference 2006, to submitted, 2006.

[9] L. Ljung, *"System Identification-Theory for the User (2nd ed.),"* Prentice-Hall, 1999.

[10] J. C. Eccles, *"Learning in the motor System,"* H. J. Freund, et al. (eds.), Progress in Brain Research, 64, Elsevier Science Pub., pp.3–18,1986.

[11] Wolpert, D.M. and Ghahramani, A., *"Computational Principles of Movement Neurosciences,"* Review, Nature Neuroscience Supplement, 3, pp.1212–1217, 2000.

[12] Toshio Furukawa and Etsujiro Shimemura, *"Predictive control for systems with time delay,"* Int. J. Control, vol.37, no.1399, pp.399–412,1983.

[13] K.Hidaka, K. Saida, S. Suzuki, *"Relation between skill level and input–output time delay,"* Int. Conference of Control Applications, pp.557–561, October 4–6, 2006.

8

Choosing the tools for Improving distant immersion and perception in a teleoperation context

Nicolas Mollet, Ryad Chellali, and Luca Brayda
TEleRobotics and Applications dept. Italian Institute of Technology
Italy

1. Introduction

The main problems we propose to address deal with the human-robots interaction and interface design, considering N teleoperators who have to control in a collaborative way M remote robots. Why is it so hard to synthetize commands from one space (humans) and to produce understandable feedbacks from another (robots) ?
Tele-operation is dealing with controlling robots to remotely intervene in unknown and/or hazardous environments. This topic is addressed since the 40s as a peer to peer (P2P) system: a single human or tele-operator controls distantly a single robot. From information exchanges point of view, classical tele-operation systems are one to one-based information streams: the human sends commands to a single robot while this last sends sensory feedbacks to a single user. The forward stream is constructed by capturing human commands and translated into robots controls. The backward stream is derived from the robots status and its sensing data to be displayed to the tele-operator. This scheme, e.g. one to one tele-operation, has evolved this last decade thanks to the advances and achievements in robotics, sensing and Virtual/Augmented Reality technologies: these last ones allow to create interfaces that manipulate information streams to synthesise artificial representations or stimulus to be displayed to users or to derive adapted controls to be sent to the robots. Following these new abilities, more complex systems having more combinations and configurations became possible. Mainly, systems supporting N tele-operators for M robots has been built to intervene after disasters or within hazardous environments. Needless to say that the consequent complexity in both interface design and interactions handling between the two groups and/or intra-groups has dramatically increased. Thus and as a fundamental consequence the one to one or old fashion teleoperation scheme must be reconsidered from both control and sensory feedback point of views: instead of having a unique bidirectional stream, we have to manage N * M bidirectional streams. One user may be able to control a set of robots, or, a group of users may share the control of a single robot or more generally, N users co-operate and share the control of M co-operating robots. To support the previous configurations, the N to M system must have strong capabilities enabling co-ordination and co-operation within three subsets: Humans, Robots, Human(s) and Robot(s).

The previous subdivision follows a homogeneity-based criteria: one use or develop the same tools to handle the aimed relationships and to carry out modern tele-operation. For instance, humans use verbal, gesture and written language to co-operate and to develop strategies and planning. This problem was largely addressed through Collaborative Environments (CE). Likely, robots use computational and numerical-based exchanges to co-operate and to co-ordinate their activities to achieve physical interactions within the remote world. For human(s)-robot(s) relationships, the problem is different: humans and robots belong to two separate sensory-motor spaces: humans issue commands in their motor space that robots must interpret, to execute the corresponding motor actions through actuators. Conversely, robots inform humans about their status, namely they produce sensing data sets to be displayed to users' sensory channels. Human-Machine Interfaces (HMI) could be seen here as spaces converters: from robot space to human space and vice versa. The key issue thus is to guarantee the bijection between the two spaces. This problem is expressed as a direct mapping for the one-to-one (1 * 1) systems. For the N * M systems, the direct mapping is inherently impossible. Indeed, when considering a 1 * M system for instance, any aim of the single user must be dispatched to the M robots. Likely, one needs to construct an understandable representation of M robots to be displayed to the single user. We can also think about the N * 1 systems: how to combine the aims of the users to derive actions the single robot must perform?

This book chapter is focused on the way we conducted researches, developments and experiments in our Lab to study bijective Humans-Robots interfaces design. We present our approach and a developed platform, with its capabilities to integrate and abstract any robot into Virtual and Augmented worlds. We then present our experiences for testing N*1, 1*M and N*M contexts, followed by two experiences which aims to measure human's visual feedback and perception, in order to design adaptative and objectively efficient N*M interfaces. Finally, we present an application of this work with a real N*M application, an actual deployment of the platform, which deals with remote artwork perception within a museum.

2. State of the art

Robots are entities being used increasingly to both extend the human senses and to perform particular tasks involving repetition, manipulation, precision. Particularly in the first case, the wide range of sensors available today allows a robot to collect several kinds of environmental data (images and sound at almost any spectral band, temperature, pressure...). Depending on the application, such data can be internally processed for achieving complete autonomy [WKGK95,LKB+07] or, in case a human intervention is required, the observed data can be analysed off-line (robots for medical imaging, [GTP+08]) or in real time (robots for surgical manipulations such as the Da Vinci Surgical System by Intuitive Surgical Inc., or [SBG+08]). An interesting characteristic of robots with real-time access is to be remotely managed by operators (Teleoperation), thus leading to the concept of Tele-robotics [UV03,EDP+06] anytime it is impossible or undesirable for the user to be where the robot is: this is the case when unaccessible or dangerous sites are to be explored, to avoid life threatening situations for humans (subterranean, submarine or space sites, buildings with excessive temperature or concentration of gas).

Research in Robotics, particularly in Teleoperation, is now considering cognitive approaches for the design of an intelligent interface between men and machines. This is because interacting with a robot or a (inherently complex) multi-robots system in a potentially unknown environment is a very high skills and concentration demanding task. Moreover, the increasing ability of robots to be equipped with many small - though useful - sensors, is demanding an effort to avoid any data flood towards a teleoperators, which would dramatically drawn the pertinent information. Clearly, sharing the tasks in a collaborative and cooperative way between all the N * M participants (humans, machines) is preferable to a classical 1 * 1 model.

Any teleoperation task is as much effective as an acceptable degree of immersion is achieved: if not, operators have distorted perception of distant world, potentially compromising the task with artefacts, such as the well know tunneling effect [Wer12]. Research has focused in making Teleoperation evolve into Telepresence [HMP00,KTBC98], where the user feels the distant environment as it would be local, up to Telexistence [Tac98], where the user is no more aware of the local environment and he is entirely projected in the distant location. For this projection to be feasible, immersion is the key feature. VR is used in a variety of disciplines and applications: its main advantage consists in providing immersive solutions to a given Human-Machine Interface (HMI): the use of 3D vision can be coupled with multi-dimensional audio and tactile or haptic feedback, thus fully exploiting the available external human senses.

A long history of common developments, where VR offers new tools for tele- operation, can be found in [ZM91][KTBC98][YC04][HMP00]. These works address techniques for better simulations, immersions, controls, simplifications, additional information, force feedbacks, abstractions and metaphors, etc. The use of VR has been strongly facilitated during the last ten years: techniques are mature, costs have been strongly reduced and computers and devices are powerful enough for real-time interactions with realistic environments. Collaborative tele-operation is also possible [MB02], because through VR more users can interact in Real-Time with the remote robots and between them. The relatively easy access to such interaction tool (generally no specific hardware/software knowledge are required), the possibility of integrating physics laws in the virtual model of objects and the interesting properties of abstracting reality make VR the optimal form of exploring imaginary or distant worlds. A proof is represented by the design of highly interactive computer games, involving more and more a VR-like interface and by VR-based simulation tools used for training in various professional fields (production, medical, military [GMG+08]).

3. A Virtual Environment as a mediator between Humans and Robots

We firstly describe an overview of our approach: the use of a Virtual Environment as an intermediate between humans and robots. Then we briefly present the platform developed in this context.

3.1 Concept

In our framework we first use a Collaborative Virtual Environment (CVE) for abstracting and standardising real robots. The CVE is a way to integrate in a standardised way of interaction heterogenous robots from different manufacturers in the same environment, with the same level of abstraction. We intend in fact to integrate robots being shipped with the related drivers and robots internally assembled together with their special-purpose operating system. By providing a unique way of interaction, any robot can be manipulated through standard interfaces and commands, and any communication can be done easily: heterogenous robots are thus standardised by the use of a CVE. An example of such an environment is depicted in Figure 1: a team of teleoperators N1;N is able to simultaneously act on a set of robots M1;M through the CVE. This implies that this environment provides a suitable interface for teleoperators, who are able to access a certain number of robots simultaneously, or on the other hand just one robot's sensor in function of the task.

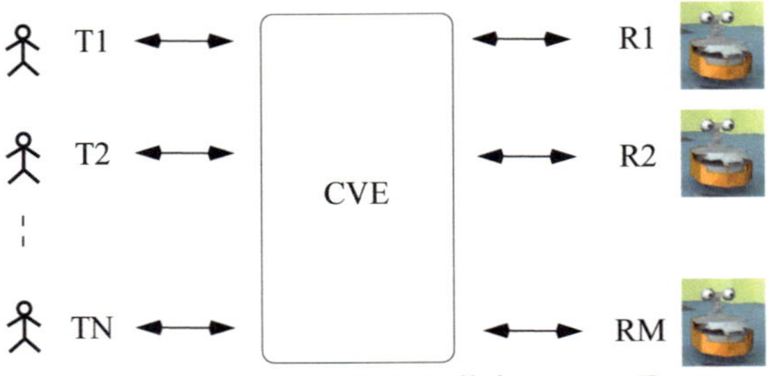

Fig. 1. Basic principle of a Virtual-Augmented Collaborative Environment: N teleoperators can interact with M robots.

3.2 Technical developments: the ViRAT platform

We are developing a multi-purposes platform, namely ViRAT (Virtual Reality for Advanced Teleoperation [MCB09][MBCF09][MBCK08]), the role of which is to allow several users to control in real time and in a collaborative and efficient way groups of heterogeneous robots from any manufacturer. We presented in the paper [MBCK08] different tools and platforms, and the choices we made to build this one. The ViRAT platform offers teleoperation tools in several contexts: VR, AR, Cognition, groups management. Virtual Reality, through its Virtual and Augmented Collaborative Environment, is used to abstract robots in a general way, from individual and simple robots to groups of complex and heterogeneous ones. Internal ViRAT's VR robots represent exactly the states and positions of the real robots, but VR offers in fact a total control on the interfaces and the representations depending on users, tasks and robots, thus innovative interfaces and metaphors have been developed. Basic group management is provided at the Group Manager Interface (GMI) Layer, through a first implementation of a Scenario Language engine[MBCF09]. The interaction with robots tends to be natural, while a form of inter-robots collaboration, and behavioral modelling, is implemented. The platform is continuously evolving to include more teleoperation modes and robots.

As we can see from the figure 2 ViRAT makes the transition between several users and groups of robots. It's designed as follows:

1. ViRAT Human Machine Interfaces provide high adaptive mechanisms to create personal and adapted interfaces. ViRAT interfaces support multiple users to operate at the same time even if the users are physically at different places. It offers innovative metaphors, GUI and integrated devices such as Joystick or HMD.

2. Set of Plug-in Modules. These modules include in particular:
   - Robot Management Module (RMM) gets information from the ViRAT interface and tracking module and then outputs simple commands to the control module.
   - Tracking Module (TM) is implemented to get current states of real environment and robots. This module also outputs current states to abstraction module.
   - Control Module (CM) gets simple or complex commands from the ViRAT interface and RMM. Then it would translates them into robots' language to send to the specific robot.
   - Advance Interaction Module (AIM) enables user to operate in the virtual environment directly and output commands to other module like RMM and CM.
3. ViRAT Engine Module is composed of a VR engine module, an abstraction module and a network module. VR engine module focuses on VR technologies such as: rendering, 3D interactions, device drivers, physics engines in VR world, etc. VR abstraction module gets the current state from the tracking module and then it abstracts the useful information, that are used by the RMM and VR Engine Module. Network Module handles communication protocols, both for users and robots.

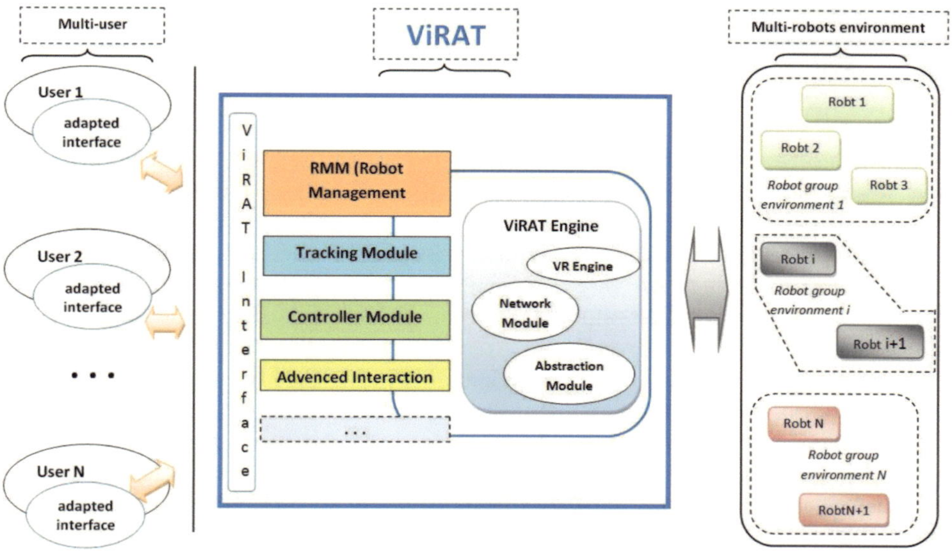

Fig. 2. ViRAT design

When a user gives some commands to ViRAT using his/her adapted interface, the standardised commands are sent to the RMM. Internal computations of this last module generate simple commands for the CM. During the running process, the TM gets the current state of the real environment and send it to the Abstraction Module, which abstracts the useful information in VIRAT's internal models of representation and abstraction. Considering this information, VR engine module updates the 3D environment presented to the user. RMM readapts its commands according to users' interactions and requests.

ViRAT project has many objectives, but if we focus on the HRI case there are two main objectives that interest us particularly for this paper:

Robot to Human

Abstract the real environment into the virtual environment: This will simplify the environment for the user. Ignorance of useless objects makes the operation process efficient. In the abstraction process, if we use a predefined virtual environment (Figure 5a), it will be initialised when the application starts running. Otherwise we construct the new virtual environment, which happens when we use ViRAT to explore an unknown area for example. After construction of a virtual environment in accordance with the real environment, we can reuse the virtual environment whenever needed. Thus the virtual environment must be adaptable to different applications. ViRAT has an independent subsystem to get the current state information from real environment termed as 'tracking module' in the previous section. The operator makes decisions based on the information perceived from the virtual environment. Because the operator does not need all the information from the tracking module, this abstraction module will optimise, abstract and represent the useful state information in real-time to user.

Human to Robot

The goal is to understand the human, and to transfer commands from the virtual environment into the real world. Several Teleoperators can interact simultaneously with 3 layers of abstraction, from the lowest to the highest (Figure 3) : the Control Layer, the Augmented Virtuality (AV) Layer, the Group Manager Interface (GMI) Layer. The Control layer is the lowest level of abstraction, where a teleoperator can take full and direct control of a robot. The purpose is to provide a precise control of sensors and actuators, including wheel motors, vision and audio system, distance estimators etc... The remaining operations, generally classified as simple, repetitive or already learnt by the robots, are executed by the Control Layer without human assistance; whether it is the case to perform them or not is delegated above, to the Augmented Virtuality Layer. Such layer offers a medium level of abstraction: teleoperators take advantage of the standardised abstracted level, can manipulate several robots with the same interface, which provides commands close to what an operator wants to do instead of how. This is achieved by presenting a Human-Machine Interface (HMI) with a purely virtual scene of the environment, where virtual robots move and act. Finally, the highest level of abstraction is offered by the Groups Manager Interface (GMI). Its role is to organise groups of robots according to a set of tasks, given a set of resources. Teleoperators communicate with the GMI, which in turns combines all the requests to adjust priorities and actions on robots through the RMM.

3.3 Goals of ViRAT

The design and tests of ViRAT allow us to claim that this platform achieves a certain number of goals:

- *Unification and Simplification*: there is a unified and simplified CVE, able to access to distant and independent rooms, which are potentially rich of details. Distant robots are parts of the same environment.

- *Standardisation*: we use a unified Virtual Environment to integrate heterogenous robots coming from different manufacturers: 3D visualisation, integration of physics laws into the 3D model, multiple devices for interaction are robot-independent.
- *Reusability*: behaviours and algorithms are robot-independent as well and built as services: their implementation is reusable on other robots.
- *Pertinence via Abstraction*: a robot can be teleoperated on three layers: it can be controlled directly (Control Layer), it can be abstracted for general commands (AV Layer), and groups of robots can be teleoperated through the GMI Layer.
- *Collaboration*: several, distant robots collaborate to achieve several tasks (exploration, video-surveillance, robot following) with one or several teleoperator(s) in real time.
- *Interactive Prototyping* can be achieved for the robots (conception, behaviours, etc.) and the simulation.
- *Advanced teleoperation interfaces*: we provided interfaces which start considering cognitive aspects (voice commands) and reach a certain degree of efficiency and time control.
- *Time and space navigation* are for the moment limited in the current version of ViRAT, but the platform is open for the next steps: teleoperators can already navigate freely in the virtual space at runtime, and will be able to replay what happened or to predict what will be (with for example trajectories planification and physics).
- *Scenario Languages applicability*. The first tests we made with our first and limited implementation of the Scenario Language for the GMI allow us to organise one whole demonstration which mixes real and virtual actors.

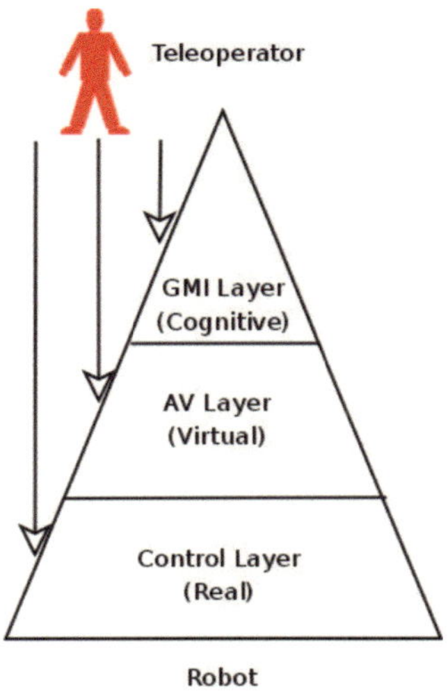

Fig. 3. In our CVE three abstraction layers (GMI, AV, Control) are available for teleoperation.

4. ViRAT's scenarios on the different actors and their interactions

As previously introduced, we aim to provide efficient N*M interfaces. To achieve such a goal, we divide the experiments in first, N*1 context, and second, 1*M context.

4.1 N*1: collaboration between humans

This basic demonstration is using one wheeled robot equipped with two cameras (figure 4). The camera video streams can be seen by a user who wear a Head Mounted Display (HMD). The movements of the HMD are tracked and transmitted to the robot's pan-tilt cameras. At any moment, this teleoperator can also see the VR world, synchronised with the real one. This VR environment is used by a second teleoperator who plan the robot's displacements. Following to this basic collaborative demonstration, we developed in the VR world a set of metaphors that for example allow to see the presence of another teleoperator in the environment, or also to understand which robot the other user is going to control, etc.

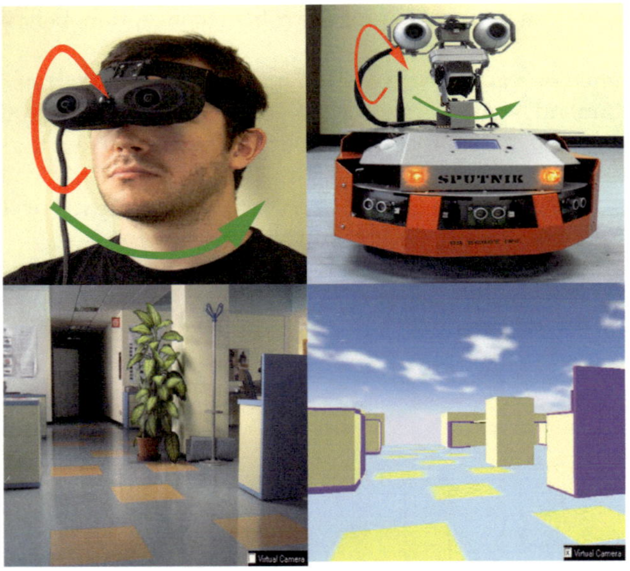

Fig. 4. One user controls robot camera, while another one uses the virtual world to choose the robot's displacements

4.2 1*M: collaboration between robots

The user supervise two robots that are collaborating to offer a real camera view to the user according to a target he pointed in the VR world (Figure 5b). One robot is a small humanoid with a camera, which moves slowly, while the second wheeled robot can go quickly to a chosen target. The user can give general commands to the robots through the Group Manager Interface, and then the RMM will generate the subtasks for this command, so it allows easily to ask to the wheeled robot to bring the humanoid one (Figure 6a), which can climb on the fast transportation robot. The TM provides the position and the orientation of the robots continuously. During the mission, user may interact with the group, showing the path and the targets to Sputnik (the wheeled robot) and redefining the requested viewpoint from the VR environment. Since HMD and Humanoid's head are synchronised, therefore user can move freely and naturally his/her head to feel immersed and present through the Humanoid's robot when this one is arrived at his final location. More details on this experiment can be found in [KZMC09].

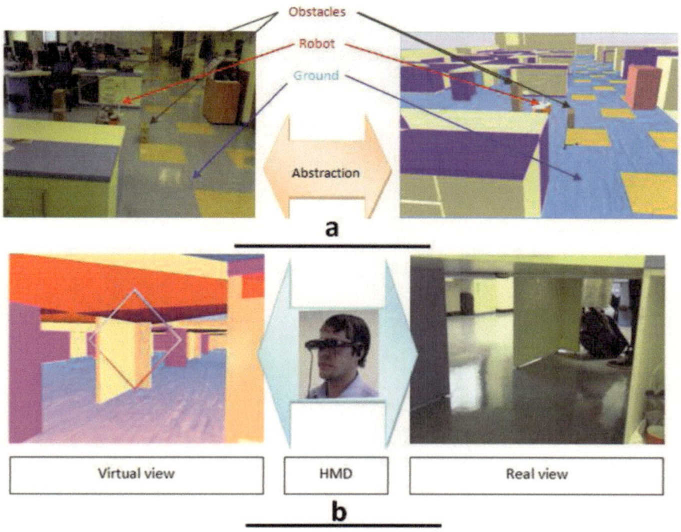

Fig. 5. VR abstraction, (a) Virtual environment interaction tool. (b) One example of some Metaphors given through VR tools

4.3 N*M: a simple case of multi-teleoperators interacting on multi-robots

We use for this demonstration two real and distant environments. The first one contains one robot, the second two robots. We considered two teleoperators who are also located in two different places. One operator acts through a PC, equipped with classical physical interfaces (mouse, keyboard, monitor), while the second one uses more immersive devices (head mounted display, joystick). The operators manage the three robots through the unified virtual world, without limitations due to site distribution. The Group Manager Interface (GMI) is responsible of scheduling and executing scenarios in function of the available resources (mainly robots' states), and to synchronise actions between the tele-operators. One of the advantages of using the VR world is that teleoperators can navigate freely in the two rooms, both when robots are moving or not. The real-time tracking of the position and velocity of the real robots (mirrored by the locations of avatars) is achieved thanks to a calibrated camera system, able to locate the real position of robots and input them in the AV Layer. The virtual avatars appear in the same virtual room, while real robots are separated in their real rooms. Thus, a distributed, complex space is represented via a unique, simple virtual room. This ViRAT's basic demonstration is a mixed of the two previous ones, so it includes all the combinations. Teleoperators can for example interact with the same robot (camera/displacements) while the GMI takes in charge the two other ones automatically.

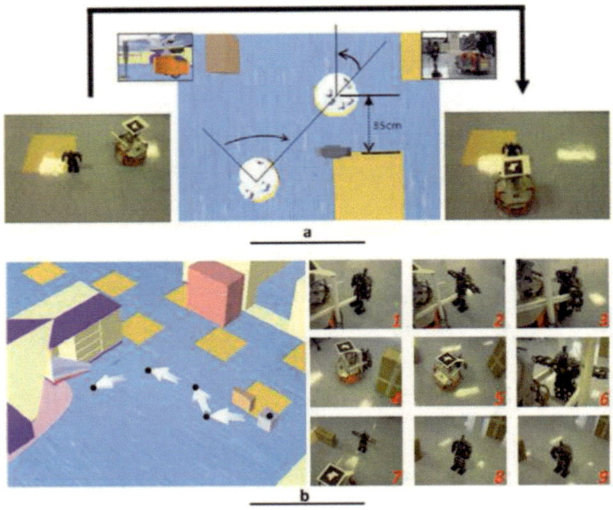

Fig. 6. Interaction through VR environment. (a) Go Near sub-task. (b) Collaboration scenario

5. Human Analysis

While we developed a set of tools for allowing the N*M general interactions pattern, we need precise analysis on human's perception in order to create adaptative and objective efficient interfaces. We present here two of our set of experiments currently in progress. We first present the influence of the height of the camera in a context of a path-to-follow task. Then, we evaluate the efficiency of a 3D map, compared to a 2D one, to allow self-localisation for the teleoperators according to a distant camera's video stream.

5.1 Height influence on task efficiency and context understanding

In this work we aim at finding ways to measure the capability of a teleoperator to achieve a simple task: a path following task that the operator must perform. The path is depicted on the ground and the user must drive the robot as close as possible to this path. The evaluation is done by comparing the path traced by the mobile robot and the original path. This allows us to drive some conclusions concerning the behaviour of the operator. Specifically, one way to measure the degree of accuracy of the task is to compute the surface between the theoretical (T) and the experimental (E) path. Each path is modelled as a curve, approximated by a piecewise linear segment joined by points in a 2D space: the approximation comes from the fact that the position and orientation of the robot is sampled, e.g by a camera acquisition system. By considering that the T and E frequently cross each other, the in-between surface S can be computed as:

$$S = \frac{1}{2} \sum_{i \in I} \sum_{p \in P_i} \begin{vmatrix} x_p x_{p+1} \\ y_p y_{p+1} \end{vmatrix} \tag{1}$$

where $I \in \{T \cap E\}$ is the set of points in which the two paths intersect, $Pi \in \{T \cup E\}$ is a subset of points between two consecutive intersections, p and p+1 are two consecutive points in each subset and x, y are the 2D coordinates of a point. The inner sum is the known Surveyor□s formula for the calculus of the area of a polygon. S can be interpreted as a surface-based error. Furthermore, because we make tests across different paths of different lengths, we can normalise by the theoretical path lengths by defining a Normalised Average Distance (NAD):

$$NAD = \frac{S}{\sum_{p \in T} \sqrt{\Delta x_p^2 + \Delta y_p^2}} \quad (2)$$

With such metric, the operators with a high/low NAD will be likely to have experienced a higher/lower deviation in following the main path. Such deviations are related to the degree of ability people have to change mentally their point of view (POV) or, on the contrary, it may represent the distortion the teleoperation system imposes to them. In other words, the deviation depends (at least partially) on the fidelity of the perception of space that each operator can feel. Figure 8(d) depicts an example of surface S, where the area is depicted in gray. The relationship is partial because other ingredients are missed such as the motor transformation between the hand actions and the robot rotations.

Experimental setup

In the experiments, the users had to follow as best as they could a stained path, by carrying out the teleoperation of an Unmanned Ground Vehicle (UGV) using a joystick for motor control output and a Head Tracking System (HTS) for perceptive control input. The users didn't have any previous knowledge about the UGV or the path, and during the experiments they could rely on the sole subjective vision by teleoperating in a separated room. To reduce the experiment variability, the speed of the vehicle was fixed to 0.15 m/s (25% of the maximum speed). This way, the user only had to care about one degree of freedom of the UGV, i.e. the steering, and two degrees of freedom for the HTS (pan & tilt): this way the comparisons can be simpler and clearer.

The experiment was carried out by 7 people (3 women and 4 men), with an age range from 22 to 46 years old. Every user made a total number of 9 trials, i.e. 3 paths by 3 POV configurations. The use of the HTS was alternated between trials, so there is an average of 3.5 users for every possible combination of paths, POV and pan & tilt. The total amount of trials is then 63 (7 users times 9 trials).

To avoid the influence between experiments, the user never made two trials in a row (the user distribution is interlaced): rather, we tried to maximize the time between two successive trials.

The scene could be observed via three different POV, each of them corresponding to a different [tilt, height] pair (see Table 5.1(a)). The height is referred to the ground level and the tilt angle is referred to the horizon: the higher the value is, the more the camera is looking down. Note that the users could not perform "self-observation", thus they were not able to develop any possible new proprioceptive model. After every trial, the users were asked to draw the shape of the path. Finally, once all trials were finished, the users filled a short form with questions regarding to the subjective perception of the experiment.

The UGV used during testing was a small vehicle (0.27m length x 0.32m width) which was built using a commercial platform. This base has four motored wheels without steering system. The speed control of each wheel is used to steer the vehicle. Figure 7(a) shows a picture of the UGV. The pan & tilt camera system was placed in a vertical guide to change the height of the camera. This system uses a manual configuration since the height was only changed between experiments and not during them. The webcam has a standard resolution of 640x480 pixels and a horizontal FOV of 36 degrees. For the experiments the frame capture was made at 15 frames per second.
The user interface is composed by three main elements:

- Head Mounted Display. The user watched the images acquired by the UGV's webcam through a HMD system (see figure 7(b)).
- Joystick. The user only controlled the steering of the UGV, since it travels at constant speed. To make the control as natural as possible, the vertical rotation axis of the joystick was chosen (see figure 7(b)). The joystick orientation was recorded during the experiments.
- Head Tracking System. To acquire the user's head movement when controlling the pan & tilt movement of the camera, a wireless inertial sensor system was used (see figure 7(b)). The head orientation was also recorded during the experiments.

During the experiments, the position and rotation of the UGV as well as the movements of the UGV's webcam were recorded at 50Hz using an optical motion capture system (Vicon). Such system acquires the position of seven markers placed on the UGV (see Figure 7(a)) by means of 10 infrared cameras (8 x 1.3Mpixel MX cameras and 2 x 2Mpixel F20 cameras). The raw data coming from this system was then properly reconstructed and filtered to extract the robot center. The user's input (joystick and HTS) was recorded with a frequency of 10Hz, since that is the rate of the UGV's commands. To analyse the data, this information was resampled to 50Hz with a linear interpolation.
Three different paths were used in the experiments because we intend to compare the results in different conditions and across different styles and path complexities. They were placed under the Vicon system, covering a surface of about 13 square meters. The first path (figure 8(a)) is characterised by merlon and sawtooth angles. The second path (figure 8(b)) has the same main shape of the first but is covered CCW by the robot and has rounded curves of different radius. The third (see figure 8(c)) is simpler with wider curves and with rounded and sharp curves. The table 5.1(b) shows a measure comparison between paths.

(a) Points of view

	1	2	3
Height (m)	0.073	0.276	0.472
Tilt angle (deg)	1.5	29.0	45.0

(b) Paths

	1	2	3
Length (m)	19.42	16.10	9.06
Width (m)	0.28	0.28	0.42

Table 1. Experimental constraints (a) and paths data (b)

Results

All the detailed results and their analysis can be found in [BOM+09]. In this work we found that performances of a basic teleoperation task are influenced by the viewpoint of the video feedback. Future work will investigate how the height and the fixed tilt of the viewpoint can be studied separately, so that the relative contribution can be derived. The metric we used allows us to distinguish between a tightly and a loosely followed path, but one limitation is that we still know little about the degree of anticipation and the degree of integration of the theoretical path that an operator can develop.

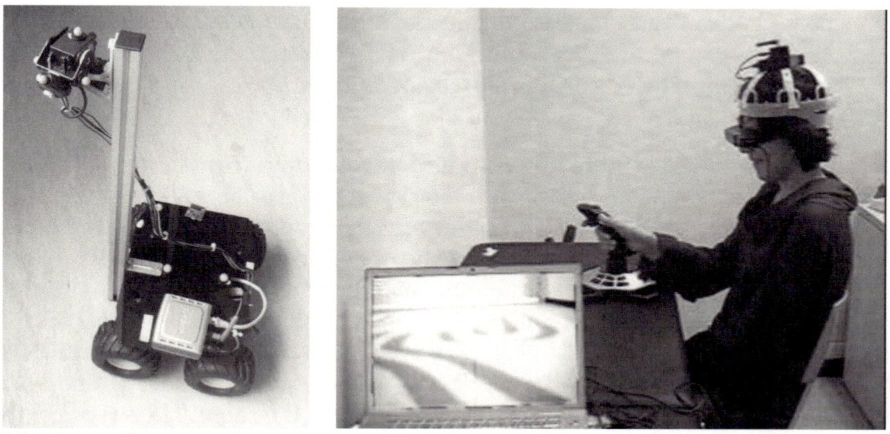

(a) General view of the UGV, with Vicon markers

(b) Operator during the experiments

Fig. 7. Experimental setup

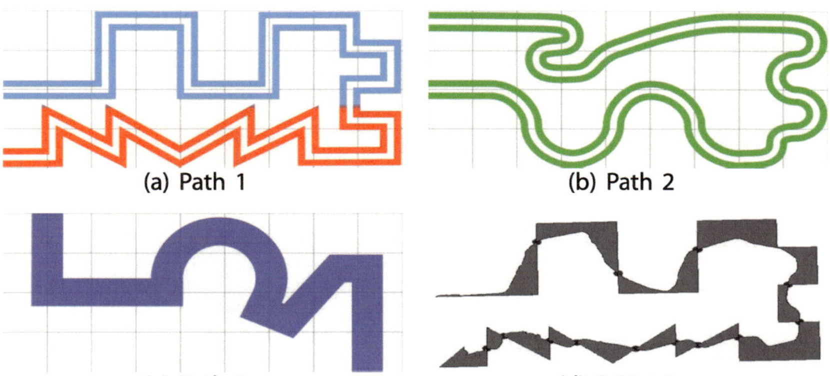

(a) Path 1

(b) Path 2

(c) Path 3

(d) S Metric

Fig. 8. Paths used for the experiments and the S metric applied to Path 1

Furthermore, we have shown that, non intuitively, the effects of a HTS were detrimental for performance: we speculate that results with an active HTS could be negative because we constrained velocity to be fixed. On one side, in fact, we added two degrees of freedom and approached a human-like behaviour, on the other side we forced the user to take decisions at an arbitrary, fixed, and as such unnatural speed, thus conflicting with the given freedom. However, we point out that the operators who positively judged an active HTS also spontaneously used the first seconds of the experiment to watch the global path, then concentrated on the requested task. The results coming from HTS could also be biased by the absence of an eye-tracking system, as the true direction of attention is not uniquely defined by the head orientation. From the questionnaire, the post-experiments drawings and from further oral comments, we can conclude that operators cannot concentrate both on following and remembering a path. This is a constraint and a precious hint for future considerations about possible multi-tasking activities. Globally speaking, our evaluations show that good performances imply that self-judgement about performance can be reliable, while the sole judgements are misleading and cannot be used as a measure of performance and no implications can be derived from them. This confirms the motivation of our study about the need of quantitative measures for teleoperation purposes.

5.2 Self representation of remote environment, localisation from 2D and 3D

The ability for teleoperators to localise remote robots is crucial: it allows them situation awareness and presence feeling and precedes any navigation or other higher level tasks. Knowing the robot's location is necessary for the operator to interact and decide about the actions to achieve safely. For some situations, mainly when the remote environment has changed or due to inherent localisation sensors uncertainty, the robot is unable to give its location neither his context. Thus, placing the robot within the tele-operator's map is meaningless. We compare here two video-based localisation methods. A tele-operator is wearing a helmet displaying a video stream coming from the robot. He can move the head freely to move the remote camera allowing him to discover the remote environment. Using a 2D map (a top view) or an interactive 3D partial model of the remote world (the user can move within the virtual representation), the tele-operator has to specify the supposed exact place of the robot.

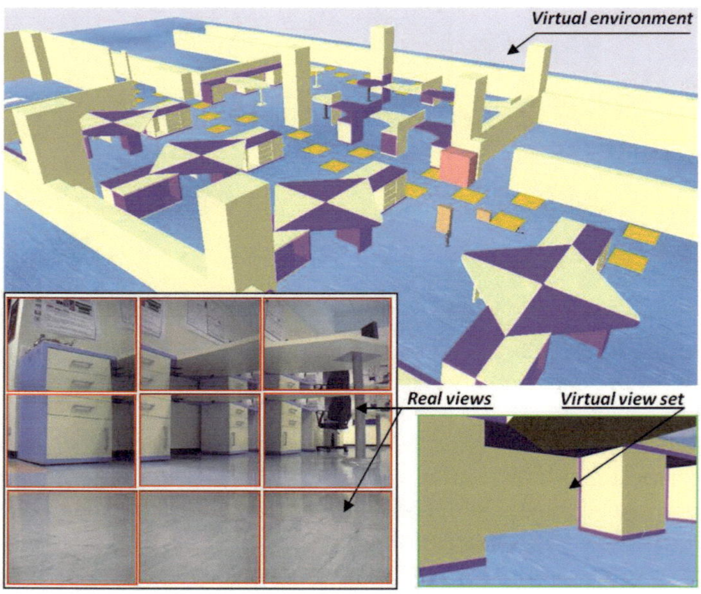

Fig. 9. 3D virtual environment used in the experiments, with a real example of robot's localization

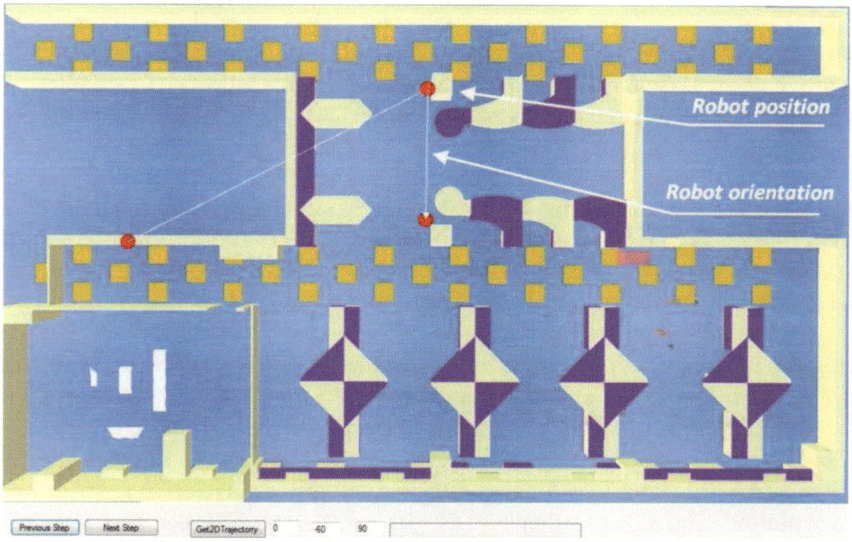

Fig. 10. 2D map used in the experiments for localization of remote robot

Experimental Setup

In 2D maps, subjects have a global view and can indicate their position and orientation on the map. In 3D display case, subjects navigate within the virtual environment till they reach their supposed position and orientation. Our laboratory was selected as the working environment for the experiments. This last contain objects of different dimensions, poses, colors and geometries such as office cabins, furniture, walls, windows, plants , etc. Practically, the robot provides the current video stream and users can move their heads with the HMD, consequently moving the robot's camera. Subjects were requested to explore the video in a minimum time and then to move to 2D maps or 3D virtual environment to localise the robot. The possibility of naturally moving the robot camera as their own head allows users to perceive information and to feel immersed in the distant location and thus to find out their location.

We have evaluated 10 subjects with 10 positions each in both 3D virtual environment and in 2D map. Thus in total the experimental scenario included 200 positions. A test session allowed the subjects to understand the meaning of the tasks and let them familiar with the 3D environment and interface.The following experimental data has been recorded: the time taken by the subject to find the robot's position, the difference between perceived and real positions in the 3D virtual environment as well as in 2D map, the perceived orientation errors with respect to the actual robot's orientation.

The first experiment aimed at finding the robot's location in the 3D environment. Subjects can navigate inside the virtual environment and then set the derived position of the robot. Figure 9 shows an example of real robot's views and the corresponding virtual view set in the 3D environment.

In the second experiment, there was only a 2D top view map. The subjects had to imagine the corresponding view and projection of the 2D points on the map to identify the view that they can see through the robot's camera. Then they pointed out the final chosen location and orientation on the 2D map (fig. 10).

Ten people of different laboratories (engineers, technicians and PhD students) have been selected as subjects for the two experiments. The subjects' age ranged from 23 to 40 years. The percentage of males was 80% and females was 20%. This variance of subjects provided a good sample space for this preliminary experimental study.

Results

Quantitative results corresponding to the 3D environment and 2D map localisation are presented in figures 11 and 12. The errors in position and orientation during the localisation of remote robot by the subjects have been noticed. As well, time (fig. 11) spent by different subjects to localise the robot has also been considered.

When using the 3D interactive environment, the average of the position error was of 48.5 cm with 2.5 degrees of orientation error. In the 2D map, the average value of position error was 100.85 cm with 5.7 degree of orientation error, so the position-orientation errors in 2D map is higher than 3D virtual environment. Possible reason for this fact is that 3D environment is richer in terms of features and landmarks subjects can rely on to derive more accurate robot position-orientation (fig. 12). In other words, the correlation between real (e.g. video data) and virtual environment representation is more effective. On the other hand, the average time consumed by all subjects in 3D was greater than 2D map. This could be due to two facts:

- the time spent in navigation in the 3D environment,
- the (quick) global view approach through the 2D top view map.

Another important result concerns personal variability: on one hand almost all subjects have the same observation concerning the time consumption and the position-orientation errors in the 3D compared to 2D. This observation could be more related to the nature of the two interfaces rather than subjects skills. On the other hand, there is a variability inter-persons: the execution time for each subject is different from others. For example, the subject number 1 has taken 117.8s to find the position in 3D and 108.5s in 2D while the subject number 7 has taken 59.8s in 3D and 30.1s in 2D.

When considering position-orientation errors, subjects' performances has been found significantly different (fig. 11b) and no correlation between 3D and 2D errors were found: subjects made big errors in 2D based localisation and perform well when using 3D environment and inversely.

The last point to notice is the distribution of the global performances in both 3D and 2D based localisation. The standard deviation is much smaller for the 3D case than for the 2D one. This suggests that the solution space in 3D is smaller than in the 2D case, and that subjects rely on the richness of the 3D environment to eliminate false estimations. This could be seen also when considering the ratio between navigation time taken by the subjects and the position error. This last in 3D is about (0.47197cm/s) is almost half of the 2D one (0.91681 cm/s). Similarly we observed that the ratio of orientation error and time consumption reduces significantly when the subjects navigate in 3D compared to 2D map.

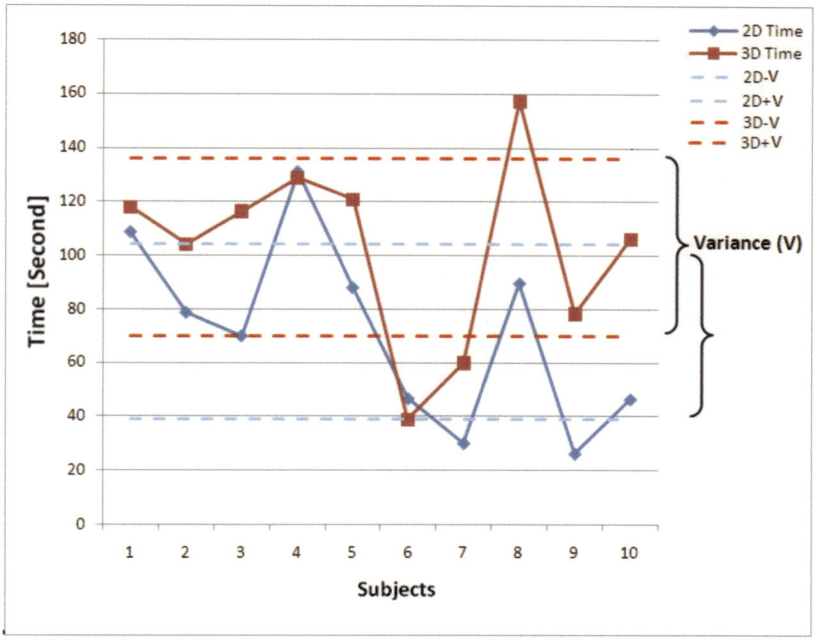

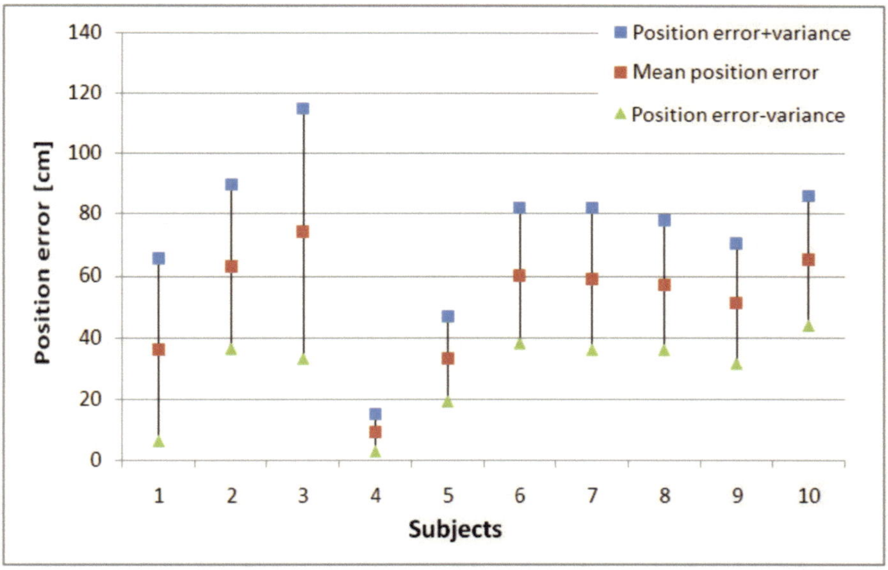

Fig. 11. (a)Average value of the time taken by subjects to find the robot. (b) Average value of position error and variance of each subject corresponding to the 3D environment interaction

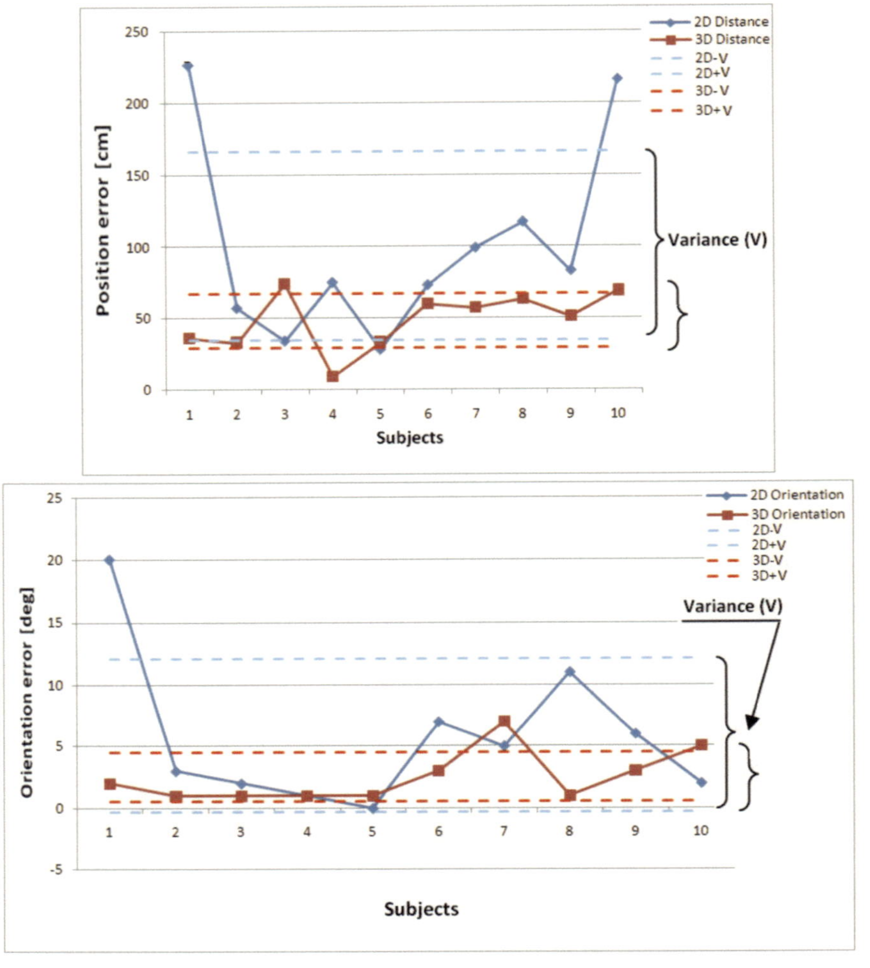

Fig. 12. (a) Average value of position error of each subject. (b) Average value of the orientation error of each subject

6. Example of a N*M real Application: improving immersion in artwork perception by mixing Telerobotics and VR

The ViRAT platform proposes now several demonstrations that are focused on interfaces or human analysis, and takes in account a set of experiments that aim to help the design of adaptative interfaces for teleoperators. We are also deploying some real-case projects with this platform. One of those is a collaboration with a museum, where the major goal is to offer the ability for distant people to visit a real museum. We'll see in this section that we are interesting in improving the sensation of immersion of real visits for virtual visitors, and that such a system may have different usages such as surveillance when the museum is closed.

The existing VR system for virtual visits of museum, like the excellent Musée du Louvre[BMCd], are still limited, with for example the lack of any natural light conditions in the Virtual Environment. Another interesting point is that the user is always alone in exploring such virtual worlds. The technologic effort to make an exploration more immersive should also take into account such human's factors: should navigation compromise with details when dealing with immersion? We believe this is the case. Does the precise observation of an artwork need the same precise observation during motion? Up to a certain degree, no. We propose a platform able to convey realistic sensation of visiting a room rich of artistic contents, while demanding the task of a more precise exploration to a virtual reality-based tool.

Fig. 13. A robot, controlled by distant users, is visiting the museum like other traditional visitors.

6.1 Deployment of the ViRAT platform

We deployed our platform according to the particularities of this application and the museum needs. Those particularities deal mainly with high-definition textures to acquire for VR, and new interfaces that are integrated to the platform. In this first deployment, consisting in a prototype which is used to test and adapt interfaces, we only had to install two wheeled robots with embedded cameras that we have developed internally (a more complete description of those robots can be found in [MC08]), and a set of cameras accessible from outside through internet (those cameras are used to track the robot, in order to match Virtual Robots locations and Real Robots locations). We modelled the 3D scene of the part of the museum where the robots are planned to evolve. A computer, where the ViRAT platform is installed, is used to control the local robots and cameras. It runs the platform, so the VR environment. From our lab, on a local computer, we launch the platform which uses internet to connect to the distant computer, robots and cameras. Once the system is ready, we can interact with the robots, and visit the museum, virtually or really.

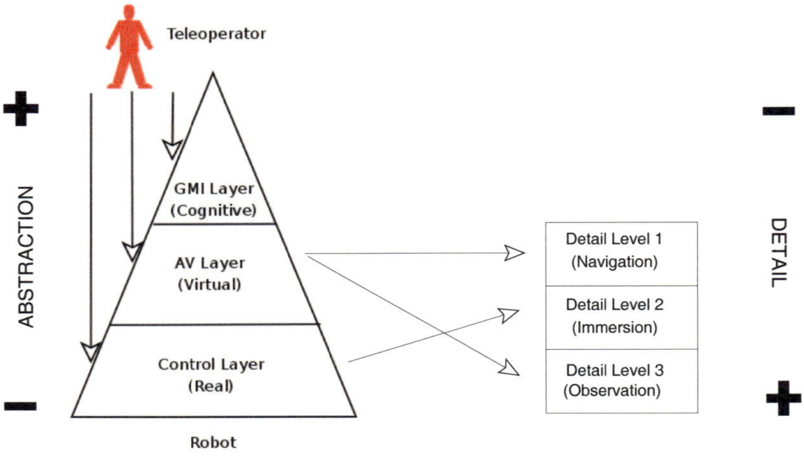

Fig. 14. Different levels of abstraction mapped into different levels of detail.

6.2 Usage of Telerobotics and VR for artwork perception

As presented in [BMCd], existing works with VR offer the ability to virtually visit a distant museum for example, but suffer from lacks of sensations: first, users are generally alone in the VR environment, and second, the degree and sensation of immersion is highly variable. The success of 3D games like «Second Life» comes from the ability to really feel the virtual world as a real world, where we can have numerous interactions, in particular in meeting other real people. Moreover, when we really visit a place, we have a certain atmosphere and ambience, which is fundamental in our perception and feeling. Visiting a very calm temple with people moving delicately, or visiting a noisy and very active market would be totally different without those feedbacks. So, populating the VR environment was one of the first main needs, especially with real humans behind those virtual entities. Secondly, even if such VR immersion gives a good sensation of presence, so of a visit, we're not really visiting the reality. Behind Second Life virtual characters, we have people sit down, in front of their computer. What about having a bijection between the reality and the virtuality ? Seeing virtual entities in the VR environment and knowing that behind those entities the reality is hidden, directly increases the feeling of really visiting, being in a place. Especially when we can switch between virtual world and real world.

Following those comments, the proposed system mixes VR and Reality in the same application. The figure 14 represents this mix, its usage, and the adaptation we made of our general framework.

On the left part, we have the degree of immersion, while on the right part, we have the level of details. The degree of immersion is made of the three levels[MBCK08]: Group Management Interface, Augmented Virtuality and Control Layer:

- First, the GMI layer, still gives the ability to control several robots. This level could be used by distant visitors, but in the actual design it is mainly used by people from the museum to take a global view on robots when needed, and to supervise what distant visitors are doing in the real museum.

- Second, the Augmented Virtuality layer, allows the user to freely navigate in the VR environment. It includes high-definition textures, coming from real high-definition photos of the art-paintings. This level offers different levels of interactions: precise control of the virtual robot and its camera (so as a consequence, the real robot will move in the same way), ability to define targets that the robot will reach autonomously, ability to fly through the 3D camera in the museum, etc.
- Third, the Control layer. At this levels, teleoperators can control directly the robots, in particular the camera previously presented. Users can see directly like if they were located at the robot's location. This level is the reality level, the users are immersed in the real distant world where they can act directly.

Fig. 15. Detail Level 1 is purely virtual, and is the equivalent of the reality (Detail Level 2)

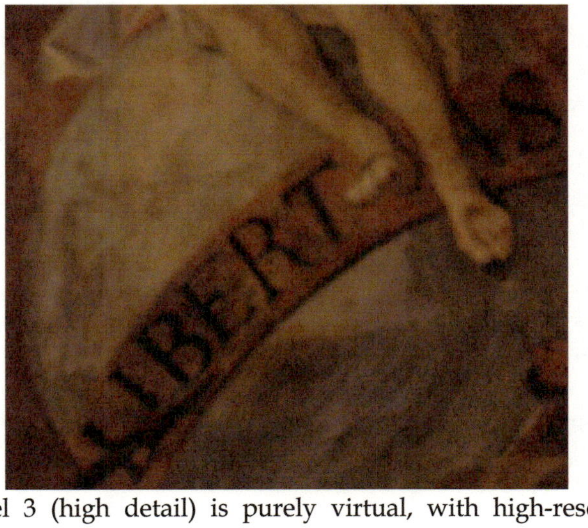

Fig. 16. Detail Level 3 (high detail) is purely virtual, with high-resolution pictures as textures. This one is used in the scene of the figure 15

On another hand, on the right part of the figure 14, the level of details represents the precision the users perceive of the environment:

- Detail Level 1 represents mainly an overview of the site and robots for navigation. The figure 15 shows the bijection between virtual and real, so the usage that a distant visitor can have of the virtual world as an abstraction of the real word.
- Detail Level 2 represents the reality, seen through the robots cameras. At this level of details, users are limited by the reality, such as obstacles and cameras limitations. But they are physically immersed in the real distant world.
- Detail Level 3 is used when distant visitors want to see very fine details of the art-paintings for example, or any art-objects that have been digitalised in high-definition. We can see in figure 16 a high-definition texture, that a user can observe in the virtual world when he wants to focus his attention on parts of the art-painting of the figure 15, that could not be accessible with the controlled robots.

When distant visitors want to have an overview of the site, and want to move easily inside, or on the opposite when they want to make a very precise observation of one art-painting for example, they use the two Detail Levels 1 and 3, in the Virtual Environment. With this AV level, they can have the feeling of visiting a populated museum, as they can see other distant visitors represented by other virtual robots, but they do not have to fit with real problems like for example occlusions of the art-painting they want to see in details due to the crowd, or displacement problems due to the same reasons. On another hand, when visitors want to feel themselves more present in the real museum, they can use the Detail level 2. This is the point where we mix Telerobotics with Virtual Reality in order to improve the immersion.

7. Conclusion

We presented in this paper our approach for designing N*M interactions pattern, and especially our objective analysis of the Human to make the interface cope with users, rather than the opposite. We presented our innovative platform, ViRAT, for an efficient teleoperation between several teleoperators and groups of robots, through adaptative interfaces. We introduced in this system our vision and usage of different levels of interactions: GMI with a scenario language, AV and direct control. We briefly presented the CVE we developed to model the robots activities and states, an environment where teleoperators can have collaborative an intermediate level of interactions with the real distant robots by using the virtual ones. We then presented in details the current experiments that are conducted to make a precise evaluation of the human's perception, to design and choose adaptative interfaces that will be objectively adapted to each teleoperator, according to contexts of tasks. We finally presented one deployment of this platform for an innovative artwork perception proposed to distant visitors of a museum. Our project is currently very active and new results come frequently. As the technical environment is ready, our actual experiments are clearly turned on human's perception evaluation, aiming the case of complex interactions with groups of robots.

We would like to make some special acknowledgments to Delphine Lefebvre, Baizid Khelifa, Zhao Li, Jesus Ortiz, Laura Taverna, Lorenzo Rossi and Julien Jenvrin for their contributions in the project and the article. The locations for our platform in the museum application are kindly provided by Palazzo Ducale, Genova.

8. References

[BMCd] L. Brayda, N. Mollet, and R. Chellali. Mixing telerobotics and virtual reality for improving immersion in artwork perception. In *Edutainment - Banff - Canada*, 2009 - to be published-.

[BOM+09] L. Brayda, J. Ortiz, N. Mollet, R. Chellali, and J.G. Fontaine. Quantitative and qualitative evaluation of vision-based teleoperation of a mobile robot. In *ICIRA 2009*, 2009.

[EDP+06] Alberto Elfes, John Dolan, Gregg Podnar, Sandra Mau, and Marcel Bergerman. Safe and efficient robotic space exploration with tele-supervised autonomous robots. In *Proceedings of the AAAI Spring Symposium*, pages 104 - 113., March 2006. to appear.

[GMG+08] Stephanie Gerbaud, Nicolas Mollet, Franck Ganier, Bruno Arnaldi, and Jacques Tisseau. Gvt: a platform to create virtual environments for procedural training. In *IEEE VR 2008*, 2008.

[GTP+08] J.M. Glasgow, G. Thomas, E. Pudenz, N. Cabrol, D. Wettergreen, and P. Coppin. Optimizing information value: Improving rover sensor data collection. *Systems, Man and Cybernetics, Part A, IEEE Transactions on*, 38(3):593–604, May 2008.

[HMP00] S. Hickey, T. Manninen, and P. Pulli. Telereality - the next step for telepresence. In *Proceedings of the World Multiconference on Systemics, Cybernetics and Informatics (VOL 3) (SCI 2000), pp 65-70, Florida.*, 2000.

[KTBC98] A. Kheddar, C. Tzafestas, P. Blazevic, and Ph. Coiffet. Fitting teleoperation and virtual reality technologies towards teleworking. 1998.

[KZMC09] B. Khelifa, L. Zhao, N. Mollet, and R. Chellali. Human multi-robots interaction with high virtual reality abstraction level. In *ICIRA 2009*, 2009.

[LKB+07] G. Lidoris, K. Klasing, A. Bauer, Tingting Xu, K. Kuhnlenz, D. Wollherr, and M. Buss. The autonomous city explorer project: aims and system overview. *Intelligent Robots and Systems, 2007. IROS 2007. IEEE/RSJ* International Conference on, pages 560–565, 29 2007-Nov. 2 2007.

[MB02] Alexandre Monferrer and David Bonyuet. Cooperative robot teleoperation through virtual reality interfaces. page 243, Los Alamitos, CA, USA, 2002. IEEE Computer Society.

[MBCF09] N. Mollet, L. Brayda, R. Chellali, and J.G. Fontaine. Virtual environments and scenario languages for advanced teleoperation of groups of real robots: Real case application. In *IARIA / ACHI 2009, Cancun*, 2009.

[MBCK08] N. Mollet, L. Brayda, R. Chellali, and B. Khelifa. Standardization and integration in robotics: case of virtual reality tools. In Cyberworlds - Hangzhou - China, 2008.

[MC08] N. Mollet and R. Chellali. Virtual and augmented reality with headtracking for efficient teleoperation of groups of robots. In *Cyberworlds - Hangzhou* - China, 2008.

[MCB09] N. Mollet, R. Chellali, and L. Brayda. Virtual and augmented reality tools for teleoperation: improving distant immersion and perception. *ToE Journal*, 5660:135–159, 2009.

[SBG+08] A. Saffiotti, M. Broxvall, M. Gritti, K. LeBlanc, R. Lundh, J. Rashid, B.S. Seo, and Y.J. Cho. The peis-ecology project: Vision and results. *Intelligent Robots and Systems, 2008. IROS 2008. IEEE/RSJ International Conference on*, pages 2329–2335, Sept. 2008.

[Tac98] S. Tachi. Real-time remote robotics-toward networked telexistence. *Computer Graphics and Applications, IEEE*, 18(6):6–9, Nov/Dec 1998.

[UV03] Tamas Urbancsek and Ferenc Vajda. Internet telerobotics for multi-agent mobile microrobot systems - a new approach. 2003.

[Wer12] M. Wertheimer. Experimentelle studien ber das sehen von bewegung,. *Zeitschrift fr Psychologie*, 61:161265, 1912.

[WKGK95] K. Warwick, I. Kelly, I. Goodhew, and D.A. Keating. Behaviour and learning in completely autonomous mobile robots. *Design and Development of Autonomous Agents, IEE Colloquium on*, pages 7/1–7/4, Nov 1995.

[YC04] Xiaoli Yang and Qing Chen. Virtual reality tools for internet-based robotic teleoperation. In DS-RT '04: *Proceedings of the 8th IEEE International Symposium on Distributed Simulation and Real-Time Applications*, pages 236–239, Washington, DC, USA, 2004. IEEE Computer Society.

[ZM91] S. Zhai and P. Milgram. A telerobotic virtual control system. In *Proceedings of SPIE, vol.1612, Cooperative Intelligent Robotics in Space II, Boston*, pages 311–320, 1991.

9

Subliminal Calibration for Machine Operation

Hiroshi Igarashi
Tokyo Denki University
Japan

1. Introduction

Although advances in computer, network and mechanical technologies promise practical applications of intelligent robots, there are still issues to be realizing available intelligence like a human. For example, global perspective recognition, long-term prediction, and experience-based intuition, which are typical human abilities, are difficult to implement in artificial systems using current technologies yet.
To overcome this problem, there are roughly two approaches. One is studies on artificial intelligence to make robots more intelligent like human beings. The other is to utilize human ability by human–machine cooperation. This paper focuses on the later.
A lot of studies on human-machine systems have been proposed. Teleoperation systems are typical application expected to improve the performance utilizing human abilities. For instance, master-slave systems provide realistic information mainly through force feedback to the operator (Horiguchi & Sawaragi, 1999) (Forsyth & Maclean, 2006). (Katsura & Ohnishi, 2007). The systems can give initiative of machine motion to human operator. However, due to constraints on machine workspace, their applications are limited to apply them to tasks fully utilizing human abilities, e.g., a situation in which unexpected disturbances or environmental changes are occurred.
Zheng et al. proposed robot teleoperation system with a mobile in disaster sites (Zheng et al., 2004). The operator's role is especially finding victims through robot vision and they focused on assisting the recognition of the victims. For mobile robot operation, teleoperation systems to assist by autonomous behaviour of the robot in response to command input are proposed (Wang & Liu, 2004) (Cheng et al., 1997). Most of these assists are based on designer's subjectivity such as giving repulsive force from obstacles, attraction from an optimum trajectory and so on. These assists were effective in a specific task.
However, such assists with external input force include two problems, namely; (i) human abilities could not be utilize, and (ii) hindering human learning ability. In description (i), the assumption that the robot knows its optimum motion could reduce the need for human involvement. Therefore, users must keep the initiative to utilize the human abilities. Furthermore, about description (ii), due to robot behaviour without operators' intention, such assists make operators not only confuse, but also hinder human learning abilities. Human learning ability is demonstrated by modifying his/her internal model of the machine motion so that the internal model closes to operated machine dynamics unconsciously (Yamada & Yamaguchi, 2004).

Proposed technique attempts to improve above two problems. In order to give an operators initiative, assist with external force by autonomous behaviour is discarded. Therefore, the machine dynamics is modified to close to operator's internal model. To estimate difference between the internal model and the machine dynamics, target tracking task, e.g. line trace, is carried out as a calibration. We expect if the internal model closed to the machine dynamics enough, the tracking error could be reduced. After the calibration, the machine has dynamics similar to the internal model and the operator could operate the machine at will.
To not hinder the human learning ability, the machine dynamics is modified without human awareness, namely "subliminal". According to cognitive science knowledge, how change of stimulus is required to notice it is quantified by "Just Noticeable Difference: JND". By considering JND, the calibration can be carried out subliminally.
Furthermore, the subliminal calibration can provide enhancement of human learning process because human learning is also to modify the internal model approaching to the operated machine dynamics (Flanagan et al., 1999). The calibration, therefore, gives high operability with short time. The subliminal calibration is implemented to vehicle operation constructed in 3D computer graphics and is verified the validity by some experiments.
This chapter is organized as follows; in the next section, a basic concept of the subliminal calibration including definition of the best operability is described. In section 3, JND by cognitive science knowledge is discussed. Next, mathematical theory of the subliminal calibration is stated in section 4. Then, in section 5, experiments and their results are presented, and finally, this article is concluded in section 6.

2. Basic Concept of Subliminal Calibration

2.1 Human initiative in human-machine systems

One of advantages in a human-machine system is that human abilities are utilized in the system behaviour. The abilities include global perspective environmental recognition, experience-based prediction, long-term planning, and so on. Therefore, a suitable human-machine system should give an operator initiative to utilize the abilities.
In order to improve the operation performance in human-machine systems, in conventional researches, most of assists involves addition of external forces to human command input based on the autonomous behaviour of the system. Such external force relies on the system designer's subjectivity, e.g. repulsive force from an obstacle. This may useful for safety in a particular task, but the assists cannot have versatility, that is, it works depending on designer's assumption. Furthermore, the assist may deprive the operator initiative by operator's unexpected motion of the system.
Further, such assist cannot consider human learning dynamics. In special, human skill can be improved through practices. A new system adapts to operator's skill, called "Human Adaptive Mechatronics: HAM" was proposed (Furuta, K., 2003)(Suzuki, S., 2005). In the HAM studies, evaluation and quantification of human skills are especially focused on. We proposed a technique that improves machine operability by taking into account human learning dynamics without adding external force to command input.

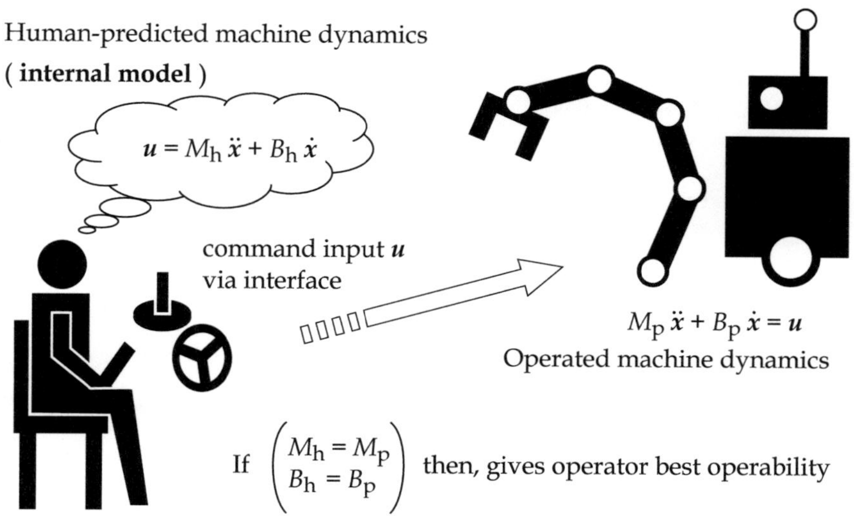

Fig. 1. Definitions of the best operability

2.2 Discussion about the best operability

Here, the best operability on machine operation systems is discussed. We defined the best operability is what a human operates the system at will. According to neural science knowledge, a human constructs in his/her brain a dynamics model of operated machine, called an internal model. As getting skill, the operator brushes up the internal model closing to the machine dynamics.

As shown in Fig. 1, if the internal model and the machine dynamics represented as machine impedance were closed to each enough, the system could give an operator the best operability. The idea of a new assist is the machine impedances modify to approaching the internal model. This is as if the machine learns the internal model like human beings.

Although the internal model should be obtained to give the assist technique during the operation, it seems to be impossible. Therefore, we implement the technique as a calibration in target tracking tasks, e.g. line trace by operated vehicle. We assume difference between the internal model and operated machine dynamics correlate with following error in the tracking task. After finishing the calibration, the machine would change the dynamics similar to the internal model, and then, the operator could operate it with high operability.

2.3 Basic concept of subliminal calibration

In our previous work, the calibration was experimented in mobile vehicle operation (H. Igarashi, 2006). The calibration, however, could not work well for the operability. One of the reasons is confusion to changing the machine behaviour according to subject's opinions.

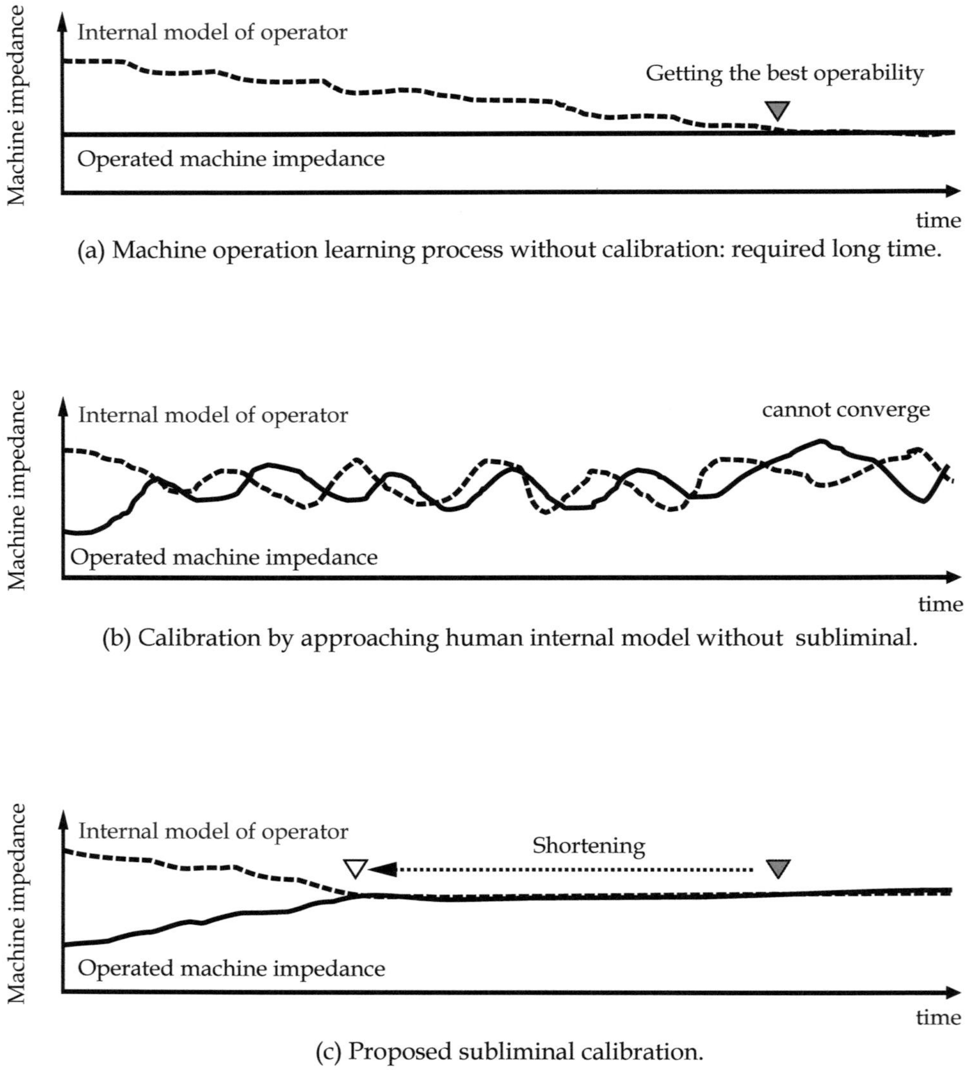

Fig. 2. Concept of proposed calibration considering human learning dynamics

This phenomenon is explained by human learning characteristics as mentioned above. That is, in spite that the operator tried to modify the internal model as shown in Fig. 2 (a), the reference model, which is represented by the machine impedances, was changed. In that case, the calibration performance was worse rather than without calibration as shown in Fig. 2 (a).
Thus, the failure gave us a new suggestion that if an operator were not aware the change of machine impedances, that is "subliminal", the learning time to get the best operability could be shorten as shown in Fig. 2 (c). About human perception characteristics, Just Noticeable

Difference: JND is investigated in cognitive science field (Teghtsoonian, 1971). In the following section, the JND to give the calibration subliminally is discussed.

3. JND for subliminal calibration

3.1 General mathematical model of operated Machine

First, for JND discussion, a simple model of operated machine in a tracking task is assumed. A motion equation of the machine with its position x and input u, is as follows;

$$\ddot{x}[k] = M[k]^{-1}(-B[k]\,\dot{x}[k] + u[k]), \tag{1}$$

where, $M[k]$ and $B[k]$ denote inertia and viscous of operated machine in time step k, respectively. These machien impedances could be variable and the calibration modifies these to approach to the human internal model. For simplicity, these are also represented as a vector;

$$\boldsymbol{\sigma}[k] = [M[k] \quad B[k]]^{\mathrm{T}}. \tag{2}$$

A following error $e[k]$ in the following task as line tracing by vehicle driving in experiments is with a target position x_r as follows;

$$e[k] = x_r[k] - x[k]. \tag{3}$$

We assume if the human internal model were enough close to operated machine impedance, the following error e could be reduced. Then, The subliminal calibration attempts to reduce $e[k+1]$ with modifying $\boldsymbol{\sigma}$[k].

3.2 Just Noticeable Difference

The JND is described that a ratio between required difference to notice the stimulus ΔI and an original stimulus I is approximately constant as follows;

$$\frac{|\Delta I|}{I} \approx c \text{ (constant)}, \tag{4}$$

where, a constant ratio c is called JND or Weber's ratio and its value depends on the kind of stimulus. For instance, that JND of pressure is 0.143 and 0.079 is for brightness was reported (Teghtsoonian, 1971). B. R. Brewer et al. applied the JND to a robotic rehabilitation system for effective rehabilitation, and investigated the JND of force and position for young and elderly subjects (Brewer at el., 2005). In our previous work, the JND for notice a difference of window size is investigated for an alert system on GUI (Igarashi et al., 2005).

Then, we consider the JND of the machine impedances is described for calibration without human awareness as following equations;

$$c_M = \frac{|M[k]-M[k-1]|}{M[k-1]}, \quad (5)$$

$$c_B = \frac{|B[k]-B[k-1]|}{B[k-1]}, \quad (6)$$

where, c_M and c_B denote JND for inertia and viscous, respectively. Hence, maximum variations the machine impedances to not notice their changes, ΔM_{max} and ΔB_{max}, are described as follows;

$$\Delta M_{\max}[k] = \begin{cases} c_M M[k-1] & \text{if } (M[k]-M[k-1]) \geq 0 \\ -c_M M[k-1] & \text{otherwise} \end{cases}, \quad (7)$$

$$\Delta B_{\max}[k] = \begin{cases} c_B B[k-1] & \text{if } (B[k]-B[k-1]) \geq 0 \\ -c_B B[k-1] & \text{otherwise} \end{cases}. \quad (8)$$

Thus, the variations of the machine impedances, ΔM[k] and ΔB[k], constraints for the subliminal calibration as following equation;

$$\Delta M[k] < M_{\max}[k], \quad (9)$$

$$\Delta B[k] < B_{\max}[k], \quad (10)$$

then,

$$\boldsymbol{\sigma}[k+1] = \boldsymbol{\sigma}[k] + \Delta\boldsymbol{\sigma}[k] = \begin{bmatrix} M[k]+\Delta M[k] \\ B[k]+\Delta B[k] \end{bmatrix}. \quad (11)$$

If the machine impedance modification were satisfied the Eqs. (9) and (10), the operator could not notice it. Finally, the subliminal impedance modification is implemented to the calibration with JND c by following filter;

$$f_{\text{JND}}(\Delta X[k], c) = \begin{cases} cX[k-1] & X[k] > cX[k-1] \\ -cX[k-1] & X[k] < -cX[k-1] \\ X[k-1] & \text{otherwise} \end{cases} \quad (12)$$

By applying this filter, the calibration with impedance modification can be conducted without operator awareness.

4. Theory of subliminal calibration

The subliminal calibration is approaching operated machine impedance to the internal model without awareness. We assume if both models were enough closed, following error in the tracking task could be reduced. In this section, theory of the calibration and its procedure are described.

4.1 Human Input Model by Neural Network

For modifying the machine impedance $\Delta\sigma[k]$ to reduce the following error $e[k+1]$, human input $u[k]$ is necessary to be predicted. The human input is predicted by the Neural Network. Input elements of the human model, $\boldsymbol{I}_{NN}[k]$, is represented as follows;

$$\boldsymbol{I}_{NN}[k] = \left[e[k] \;\; \cdots \;\; e[k-N_e] \;\; u[k] \;\; \cdots \;\; u[k-N_u]\right], \tag{13}$$

where, N_e and N_u denote the number of tracing steps of the following errors e and input u, respectively. Thus, predicted input $\hat{u}[k+1]$ is estimated as;

$$\hat{u}[k+1] = f_{NN}(\boldsymbol{I}_{NN}[k]), \tag{14}$$

where, $f_{NN}(\bullet)$ represents a forwarding Neural Network which is carried out back propergation learning with $\boldsymbol{I}_{NN}[k-1]$ and its teaching signal $u[k]$ in real-time.

4.2 Subliminal modification of machine impedances

Variation of the machine impedances $\Delta\boldsymbol{\sigma}$ to reduce following error e is defined here.
First, an evaluation function of following performance $J[k]$ using predicted input is calculated as follows;

$$J[k] = \sum_{i=k}^{k+T_p} (\mu_e\, \hat{e}[i]^2 + \mu_0 \left|\sigma[0] - \sigma[i]\right|^2, \tag{15}$$

where,

$$\hat{e}[i] := (x_r - \hat{x}[i]), \tag{16}$$

$$\dot{\hat{x}}[k] := \dot{x}[k], \tag{17}$$

$$\hat{x}[k] := x[k], \tag{18}$$

$$\ddot{\hat{x}}[i] = M[i]^{-1}(-B[i]\dot{\hat{x}}[i] + \hat{u}[i]) \quad i = k, \cdots, k+T_p. \tag{19}$$

The evaluation value $J[k]$ is determined with estimated motion of the machine until T_p step later by using predicted input $\hat{u}[k]$, and T_p denotes time step number for predictive virtual motion. Then, μ_e is weights for the evaluation, and μ_0 represents a weighting coefficients to constrain of impedance parameter divergence with initial impedance $\boldsymbol{\sigma}[0]$.
Finally, the variation of impedance $\Delta\sigma[k]$ by steepest descent method with JND filtering in Eq. (12) is as follows;

$$\Delta\sigma[k] = -f_{JND}\left(\frac{\partial J[k]}{\partial \sigma}, \boldsymbol{c}\right), \tag{20}$$

where,

$$c := \begin{bmatrix} c_M \\ c_B \end{bmatrix}. \tag{21}$$

By Eq. (20), a calibration without human awareness is realized for machine operation system. After the subliminal calibration, the impedances $\boldsymbol{\sigma}[k]$ are fixed to give high operability to the operator.

Note that, human would operate the machine at will soon even if differences between the internal model and the machine impedance $\boldsymbol{\sigma}[k]$ are remains because human has a learning ability as swhon in Fig 2 (a). In other words, this technique enhances the human learing dynamics rather than operation performance.

5. Experiments and Results

Vehicle driving with the subliminal calibration is experimented (see Fig. 3). The vehicle model is based on hovercraft model because of difficult to drive and required getting skill for high performance.

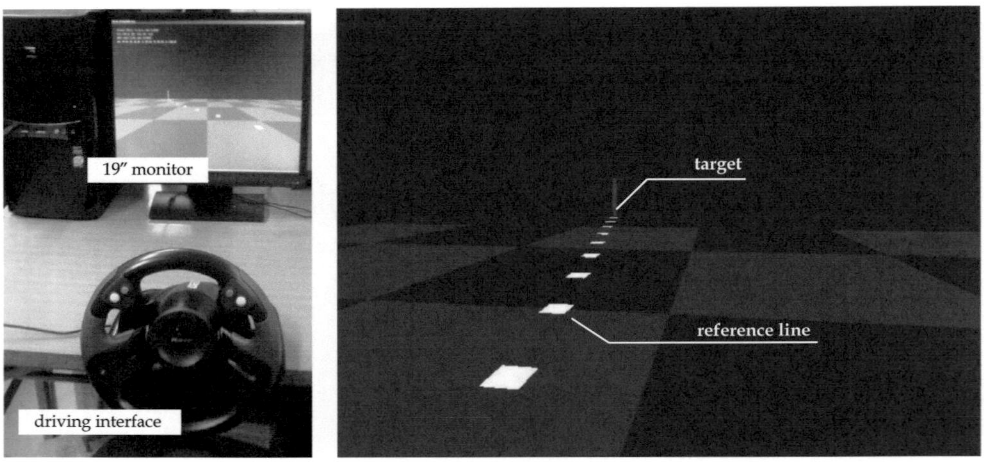

(a) Experiment setup. (b) Operator view in line tracking experiment.

Fig. 3. View of experimental environment in line trace task

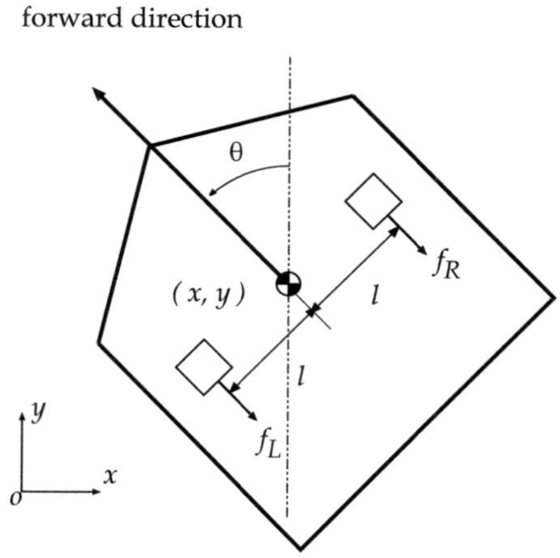

Fig. 4. Model of operated vechile in the experiment

5.1 Operated vehicle model

The equations of motion for an operated vehicle, which is with hovercraft-normative dynamics as shown Fig. 3, is described here. Where (x, y) denotes the global positioning axes of the vehicle, and θ the angle to forwarding direction.

Equations of motion are expressedas as follows, where $M[k]$ denotes mass, $B[k]$ translational viscosity, $I[k]$ inertial moment, and $N[k]$ rotational viscosity:

$$\ddot{x}[k]=\frac{1}{M[k]}\{-B[k]\dot{x}[k]+\cos\theta[k](f_L[k]+f_R[k])\}, \tag{22}$$

$$\ddot{y}[k]=\frac{1}{M[k]}\{-B[k]\dot{y}[k]+\sin\theta[k](f_L[k]+f_R[k])\}, \tag{23}$$

$$\ddot{\theta}[k]=-\frac{N[k]}{I[k]}\dot{\theta}[k]+\frac{l}{4I[k]}(f_L[k]-f_R[k]). \tag{24}$$

Where, f_L[k] and f_R[k] denote output of thrusters as;

$$f_L[k]:=\alpha\, u_a[k](1-u_s[k]), \tag{25}$$

$$f_R[k]:=\alpha\, u_a[k](1+u_s[k]). \tag{26}$$

where, u_a [k] ($-1 \le u_a$ [k] ≤ 1) denotes command input for forward movement, u_s [k] ($-1 \le u_s$ [k] ≤ 1) rotation, l distance between thrusters, and α a constant.

The subliminal calibration modifies M, B, I, and N closer to the internal model by reducing following error. The machine impedances are rewriten Eq. (12) into following vector;

$$\sigma[k]=\begin{bmatrix}M[k] & B[k] & I[k] & N[k]\end{bmatrix}^T. \tag{27}$$

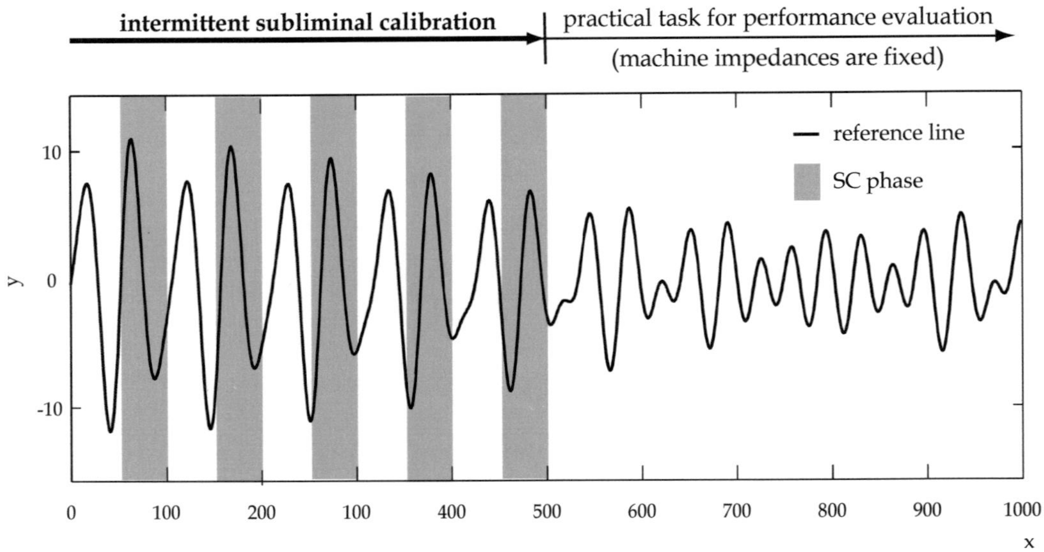

Fig. 5. Driving course and area of giving the Subliminal Calibration (SC) in the experiment

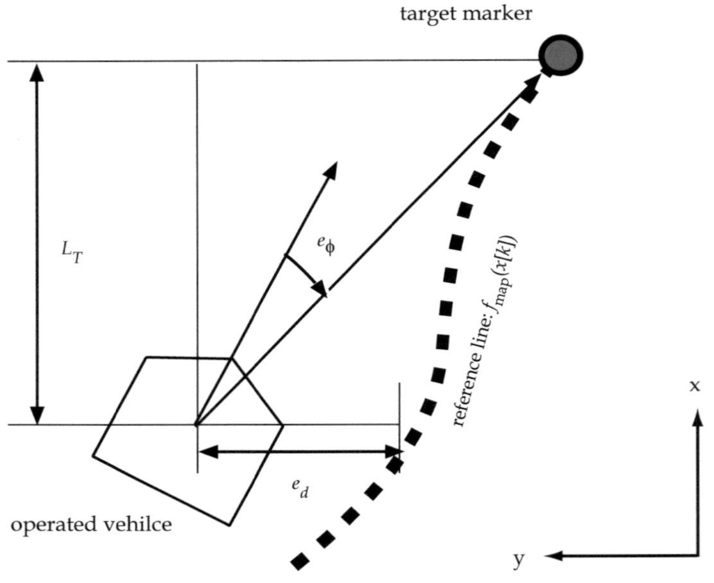

Fig. 6. Evaluation indexes for vehicle operation experiment

Similar to Eqs. (5) and (6), c_M, c_B, c_I and c_N are represented as JNDs on M, B, I, and N, respectively. According to our prior experimental results, $c_M = c_B = c_I = c_N = 0.05$ are small enough for subliminal calibration for the vehicle operation.

As shown in Fig. 4, participants use a driving interface to operate the vehicle in the OpenGL 3D-CG environment. In this experiment, the accelerator was set at a constant $u_a[k] = 1.0$ to focus on sterring performance.

5.2 Experimental Setup

The line tracking task was to have participants maneuver a vehicle through a commercial driving interface (R220, Saitek Ltd.) while observing robot camera views displayed on a 19-inch monitor. Participants were instructed to follow the reference line as closely as possible by steering only with accelerator input set constant. The driving performance, with/without calibration, for 8 participants was evaluated. To reduce effects on unconscious learning, subjects had breaks between two trials, and experiments were conducted in a random sequence.
To prevent participants from memorizing the reference line, the line generation was expressed as a combination of sine waves shown in Fig. 5;

$$f_{\mathrm{map}}(x) = 5\sin 0.11x + 4\sin 0.07x + 3\sin 0.05x\,. \tag{30}$$

The goal of the course is set at $x = 1000$ and it takes about three minutes to finish. In this task, the subliminal calibration was set for $x < 500$ and performances with/without it was compared. The calibration was conducted intermittently as shown in Fig. 6, to let participants display adaptability and learning unconscioully.
Performance evaluation indexes of the line tracing are shown in Fig. 6, following error on the Y-axis from the reference line is expressed by e_d and those in orientation between the target and the the vehicle, by e_ϕ, then we obtain the following evaluation values:

$$e_d[k] = y[k] - f_{\mathrm{map}}(x[k]), \tag{28}$$

$$e_\phi[k] = \tan^{-1}\frac{f_{\mathrm{map}}(x[k]+L_T) - y[k]}{L_T} - \theta[k]\,, \tag{29}$$

Where, L_T denotes distance on the x-axis from the vehicle to a target flag, which moves on reference line; with $L_T = 10$ in the experiment.
The human model is constructed based on these evaluation values, namely input vector to the Nerural Network in Eq. (13) is rewriten as follows;

$$\boldsymbol{I}_{NN}[k] = \lfloor e_d[k] \;\cdots\; e_d[k-N_e] \;\; e_\phi[k] \;\cdots\; e_\phi[k-N_e] \;\; u_s[k] \;\cdots\; u_s[k-N_u] \rfloor\,. \tag{30}$$

Where, in the experiments, $N_e = N_u = 5$.
Because accelalation input $u_a = 1.0$, human input is only predicted for sterring input u_s. Therefore, using forwarding Neural Network f_{NN} with input $\boldsymbol{I}_{NN}[k]$ in Eq. (30) , the input prediction is as;

$$\hat{u}_s[k+1] = f_{NN}(\boldsymbol{I}_{NN}[k])\,. \tag{31}$$

Then, the modification of the impedance is as follows;

$$J[k] = \sum_{i=k}^{k+T_p} (\mu_d \hat{e}_d[i]^2 + \mu_\phi \hat{e}_\phi[i]^2 + \mu_0 \left|\sigma[0] - \sigma[i]\right|^2 , \quad (32)$$

wehre, the initial value of the machine impedance was set to $\boldsymbol{\sigma}[0]$ = $[50, 50, 50, 50]^T$. In experiments with the calibration, initial $\boldsymbol{\sigma}$ would be modified to minimize $J[k]$ by Eq. (20).

5.3 Experimental Results

Fig. 7 shows averages of line tracking performance for all 8 participants. All participants couuld NOT notice changes of the machine impedance by the subliminal calibration through the experiment. Fig. 7 (a), (b), (c) and (d) show line following evaluation without the subliminal calibration. Note that, inspete of without the calibation, following errors decreased as x increased, apparently representing improvements due to operator learning dynamics. This indicates that participants can have their internal model approach operated vehicle dynamics even without the calibration. Fig. 7 (e), (f), (g) and (h) show results of giving subliminal calibration intermittently in $x < 500$.

By Fig. (e) and (g) following error e_d and e_ϕ improved over the case without the calibration, and standard deviation also improved as shown in Fig. 7(f) and (h). Note that machine impedances were intermittently updated in calibration of $x < 500$ but no updating was done for $x > 500$. This suggests that the calibration in the first half of the course has transformed the machine whoes impedance into the internal model and the operator could get high operability.

Next, Fig. 8 shows typical two participants who are the most skilled and unskilled operator are focused on. According to Fig. 8 (a), the skilled operator is enough high performance without the calibration. Note Fig. 8 (b) that, the subliminal calibration did not interfere in the skilled operator.

Fig. 8 (c) and (d) shows the following performance of an unskilled operator. As shown in Fig. 8 (d), even after the calibration x > 500, the unskilled operator could keep performance, that is, the machine impedance could approach to the internal model of the unskilled operator. Therefore, the subliminal calibration can be applied without concern of operator's skill.

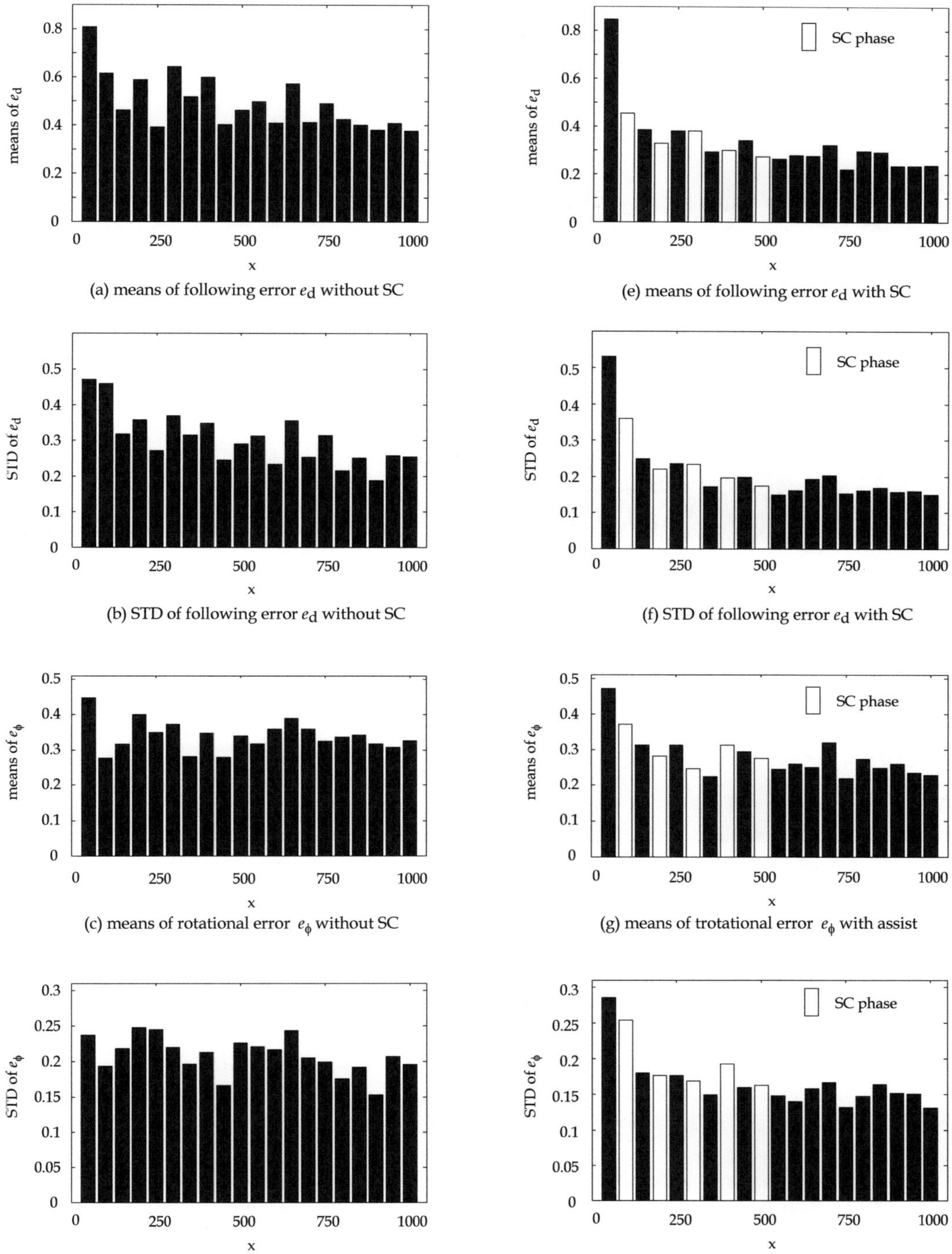

(a) means of following error e_d without SC

(e) means of following error e_d with SC

(b) STD of following error e_d without SC

(f) STD of following error e_d with SC

(c) means of rotational error e_ϕ without SC

(g) means of trotational error e_ϕ with assist

(d) STD of rotational error e_ϕ without SC

(h) STD of rotational error $e_\phi$with SC

Fig. 7. Experimental results: means and STD of the evaluation value e_d and e_ϕ with/without the Subliminal Calibration (SC).

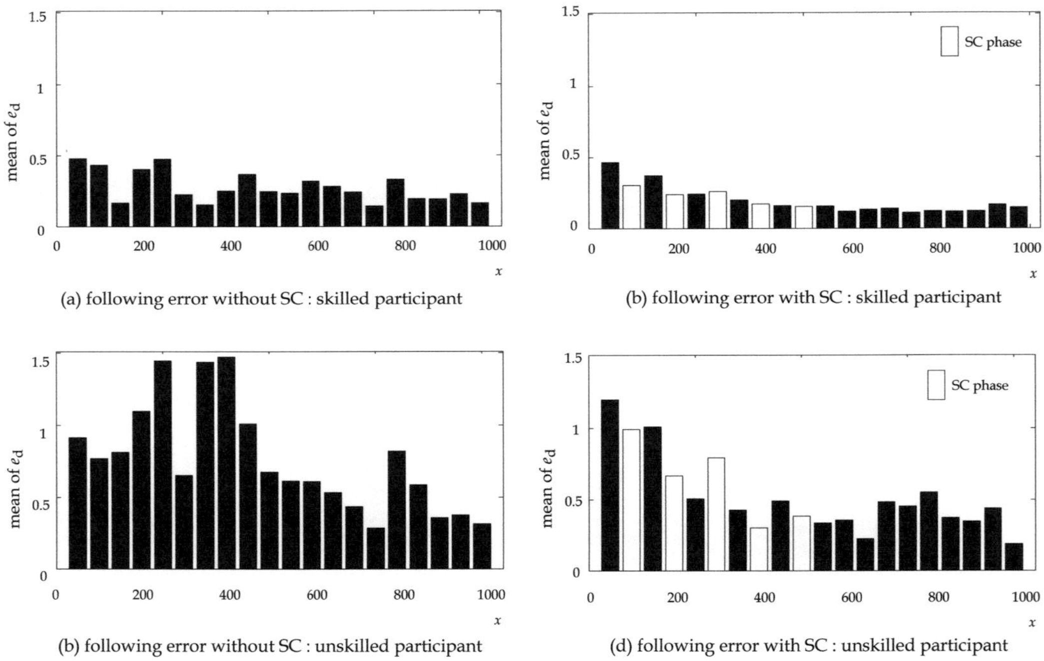

Fig. 8. Experiment results: following performance in skilled and unskilled operator.

6. Conclusion

Most of conventional assist techniques for human-machine systems relied on autonomous behaviour of the system with adding external force to command input, and thus on the subjectivity of system designers. These techniques, however, may work well only in a specific task a system designer assumed, and also may hinder the human initiative and learning ability.

Therefore, we proposed a calibration technique approaching the machine impedance to operator's internal model. This is expected to maximize the operability with which operator maneuver a machine at will with maintaining the initiative.

Human learning ability makes them feel uncomfortable in the face of variations in machine dynamics and it brings for the worse of operability. We set up a criterion on varying the operated machine impedance using JND as a perception criterion for varying stimulation. Variations in machine impedance within the limits of the criterion ensure calibration subliminally.

Our proposal requires estimation of an operator's internal model, so we conducted calibrations during a line tracking task in the experiment. The results show that all subjects improved line following accuracy without being aware of variations in operated vehicle impedance. This means that the vehicle dynamics approached the internal model, confirming that accuracy of $x > 500$ was maintained even after the calibration stopped. In other words, our proposed calibration effectively customizes operated machine dynamics for individual subjects.

Acknowledgement

This research was partially supported by the Ministry of Education, Science, Sports and Culture, Grant-in-Aid for Scientific Research (A) (No. 20246071).

7. References

Brewer, B.; Fagan, M. ; Klatzky, R. & Matsuoka, Y. (2005). Perceptual Limits for a Robotic Rehabilitation Environment Using Visual Feedback Distortion. *IEEE Transactions on Neural Systems and Rehabilitation Engineering*, Vol. 13, No. 1, pp.1-11

Cheng, G.; Zelinsky, A. & Zelinsky, E. (1997). Supervised Autonomy: A Paradigm for Teleoperating Mobile Robots. *Proceedings of the IEEE/RSJ International Conference on Intelligent Robots and Systems*, pp. 973-980

Flanagan, J.; Nakano, E.; Imamizu, H.; Osu, R.; Yoshioka, T. & Kawato, M. (1999). Composition and decomposition of internal models in motor learning under altered kinematic and dynamic environments. *Journal of Neuroscience*, 19:RC34

Forsyth, B. & Maclean, K. (2006). Predictive Haptic Guidance: Intelligent User Assistance for the Control of Dynamic Tasks. *IEEE Transactions on Visualization and Computer Graphics*, Vol. 12, No. 1, pp.103-113

Furuta, K. (2003). Control of Pendulum: From Super Mechano-System to Human Adaptive Mechatronics. *Proceedings of the 42nd IEEE Conference on Decision and Control*, plenary talk, pp. 1498-1507

Horiguchi, Y. & Sawaragi, T. (1999). Human-Robot Collaboration within a Virtual Space Using Behavioral Task-Morphologies. *Proceedings of International Conference of IEEE System, Man and Cybernetics*, Vol. 1, pp.756-761

Igarashi, H.; Takeya, A.; Kubo, Y.; Suzuki, S.; Harashima, F. & Kakikura, M. (2005). Human Adaptive GUI Design for Teleoperation System. *Proceedings of the 31st Annual Conference of the IEEEE Industrial Electronics Society*, pp.1973-1978

Igarashi, H.; Takeya, A.; Harashima, F. & Kakikura, M. (2006). Human Adaptive Assist Planning for Teleoperation. *Proceedings of The 32nd Annual Conference of the IEEE Industrial Electronics Society*, pp.4522-4527

Katsura, S. & Ohnishi, K. (2007). "Acquisition and Analysis of Finger Motions by SkillPreservation System, *IEEE Transactions on Industrial Electronics*, Vol. 54, No. 6, pp.3353-3361.

Suzuki, S.; Harashima, F. & Pan, Y. (2005). Assistant control design of HAM including human voluntary motion. *Proceedings of the 2nd COE Workshop on Human Adaptive Mechatronics*, pp.105-110

Teghtsoonian, R. (1971). On the Exponents in Sevens's Law and the Constant in Ehrman's Law. *Psychological Review*, Vol. 78, pp. 71-80

Wang, M. & Liu, J. (2004). A Novel Teleoperation Paradigm for Human-Robot Interaction. *Proceedings of IEEE International Conference on Robotics, Automation and Mechatronics*, pp. 13-18

Yamada, S. & Yamaguchi, T. (2004). Training AIBO like a dog, *Proceedings of the 13th International Workshop on Robot and Human Interactive Communication*, pp.431-436

Zheng, X.; Tsuchiya, K.; Sawaragi, T.; Osuka, K.;Tsujita, K.; Horiguchi, Y. & Aoi, S. (2004). Development of Human-Machine Interface in Disaster-Purposed Search Robot Systems That Serve as Surrogates for Human. Proceedings of IEEE International Conference on Robotics and Automation, pp. 225-230

10

Cable driven devices for telemanipulation

Carlo Ferraresi and Francesco Pescarmona
Department of Mechanics - Politecnico di Torino
Italy

1. Introduction

One of the most interesting developing fields in modern telemanipulation research is the use of a slave robot commanded by a kinematically different master. The interest in asymmetrical master-slave telemanipulators arises from the desire to design a master which will be as efficient as possible for the operator, whereas the symmetric arrangement would constrain the master design to the same strict requirements of the slave one.

The hand controllers (or joysticks) are among the most effective means for human operators to control the complex machines used in telemanipulation systems. From the point of view of the man-machine interface they can be seen as motion input devices, because the control system reads their sensors to elaborate the trajectories to be imposed to the remote manipulator.

An important characteristic of telemanipulators is the possibility to make the operators feel like they were at the remote site, actually performing the manipulation task. It is well known that not only position feedback, but also force feedback from the remote machine to the human operator is necessary to obtain good performance of telemanipulation in training simulators or in hazardous environment operations, providing the sense of balance and the feeling of touching real objects (McAffee & Fiorini, 1991; Conklin & Tosunoglu, 1996; Batsomboon et al., 2000), thus realising a *haptic* device (haptics is the science that deals with the sense of touch).

Force feedback hand controllers are actuated by motors, so that the control system can exert forces on the operator's hand; they can be seen as devices that output forces, therefore sometimes they are referred to as *force displays*.

Most of the existing devices have serial or parallel mechanical architectures. In a serial structure, the necessity to move most (or all) of the actuators tends to add weight and inertial forces to be controlled, conflicting with any force feedback received from the remote manipulator.

Even with the use of counterbalancing weights to equilibrate the mass effects of the structures, serial controllers still experience their intrinsic drawbacks. The advantages are greater than the disadvantages when a serial structure is used as a manipulator (large workspace, simple joint position control, etc.), but, when implemented as a manual master controller, their size often becomes too large, and their weight too heavy for practical use.

Moreover the play at joints and links and the errors at each joint variable measurement increase as one moves towards the payload, which implies bad precision at the end effector.

A parallel structure, on the other hand, usually allows placing all motors, brakes, and accessories at one centralized location. This eliminates the necessity to carry and move most of the actuators as happens in the serial case. Hence, input power is mostly used to support the payload, which is approximately equally distributed on all the links; the stress in the link is mostly traction-compression which is very suitable for linear actuators as well as for the rigidity, then excellent load/weight ratios may be obtained.

Parallel link mechanisms also present other interesting features: the position of the end effector of a parallel manipulator is much less sensitive to the error on the articulation sensors than for serial link robots. Furthermore, their high stiffness ensures that the deformations of the links will be minimal and this feature greatly contributes to the high positioning accuracy of the end effector.

A class among parallel devices, cable robots are parallel devices using cables as links. They have been proposed for the realization of high speed robot positioning systems needed in modern assembly operations (Kawamura et al., 1995).

Cable-actuated parallel devices represent an interesting perspective. They allow great manoeuvrability, thanks to a reduced mass, and also promise lower costs with respect to traditional actuators. Furthermore, the stroke length of each linear joint does not follow the same restrictions as with conventional structures (pantograph links, screw jacks, linear actuators), because cables can be extended to much higher lengths, for instance unwinding from a spool. This feature allows achieving the advantages typical of parallel mechanical structures without particular requirements on the positioning of motors, brakes, sensors and other accessories, giving the possibility to optimise the ratio between the device workspace volume and its total size.

On the other side, this type of actuation is totally irreversible (cables can only be pulled by the motors and they obviously cannot push). Therefore, to get a six degree of freedom device, it is necessary to have at least seven forces acting on the end effector. On Earth, gravity on the moving part exerts a constant force which may be considered in the force closure calculation. Therefore, six cables are sufficient for specific applications where no acceleration higher than g is required, at least downwards. Several examples exist of this kind of device, e.g. cranes (Bostelman et al., 1996). However, normally, higher accelerations are needed; therefore, most applications need at least seven cables with the corresponding actuators.

Cable-driven devices can be also employed as force feedback hand controllers, fixing on a handle several cables stretched by motors and leaning over pulleys, to effect force reflection; the measurement of the cable lengths allows obtaining position and orientation of the handle, determining the kinematical variables to be sent to the control system of the slave arm. Moreover, composing the traction forces of the cables, a six-dimensional wrench can be exerted on the operator's hand, representing the reactions acting on the slave robot.

On the other hand, the use of lightweight cables might induce undesired vibrations which could disturb the operator, overlapping the force feedback; therefore the necessary actuator redundancy may also be exploited to increase the device stiffness, producing suitable internal forces, contributing to a higher positioning accuracy of the manipulator as well.

Furthermore, cable redundancy is also useful to overcome another disadvantage typical of parallel mechanisms: the forward kinematics problem is not simple and, generally, many solutions for every actuator configuration are obtained, among which it is not always possible to distinguish the correct one actually reached by the end effector: in this case the

redundancy will help in the exclusion of solutions which may appear mathematically possible but do not correspond to reality. Finally, the number of cables greatly influences shape and size of the workspace and the overall device dexterity.
This chapter deals with the main peculiar aspects that must be considered when developing a cable-driven haptic device, with particular regard to the algorithms for geometric, kinematical and static analysis, to the control system and to the mechanical aspects typical of this kind of application.

2. Geometry

As pointed out in Section 1, designing a cable driven device with n degrees of freedom requires at least $n+1$ cables in a convenient layout. Apart from particular cases, it is often interesting to be able to control six degrees of freedom ($n = 6$); therefore, in the following structures with at least seven cables will be considered.
The conceptual scheme of a cable driven device is shown in figure 1.

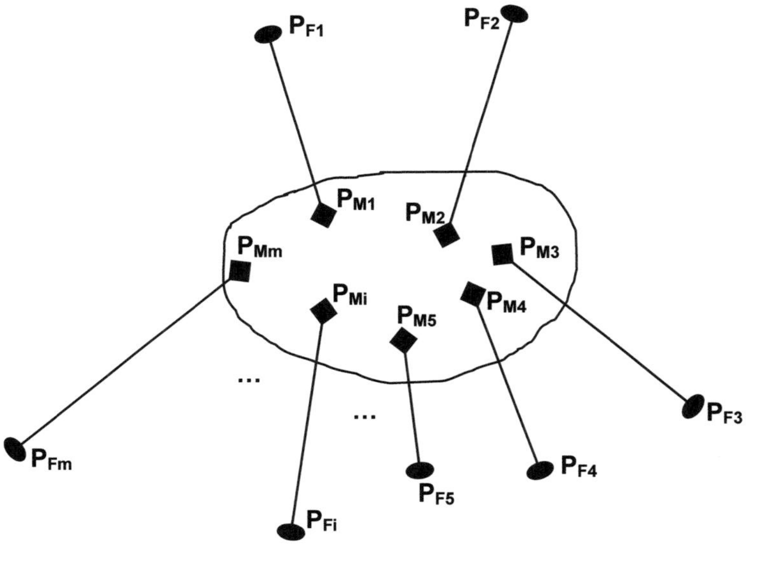

Fig. 1. Generic scheme of a cable driven device.

The moving part is a solid body of any shape, carrying the end effector or the handle for the user to operate. A total number of m cables are attached by one end to it. The point in which the i-th cable is attached to the moving part is called P_{Mi}. Towards the other end, each cable passes through a guide such as a bored support or a pulley, which conveys it to a spool, linear motor or whatever mechanism allows its motion.
For geometric purposes, it is convenient that the guide through which the cable passes is made in such a way that it is possible to identify a single, fixed point called P_{Fi} where the cable passes in all of its configurations. This way, the remaining part of the cable holds no

interest and, simplifying, each cable can be treated as an actuator of variable length attached to the fixed frame in the point P_{Fi} and to the moving part in the point P_{Mi}.
Apart from particular cases, it is convenient to design a well-organized layout of fixed and moving points to simplify geometry, kinematics and most of all control of the device. For instance, having the points lay on planes or making two or more cables converge to a single point can lead to significant simplifications, as will be pointed out later in this Section.
As examples, consider the two structures in figure 2. Note the presence of a fixed frame, referred to as *base*, while the moving part, connected to one end of each cable, is called *platform*.

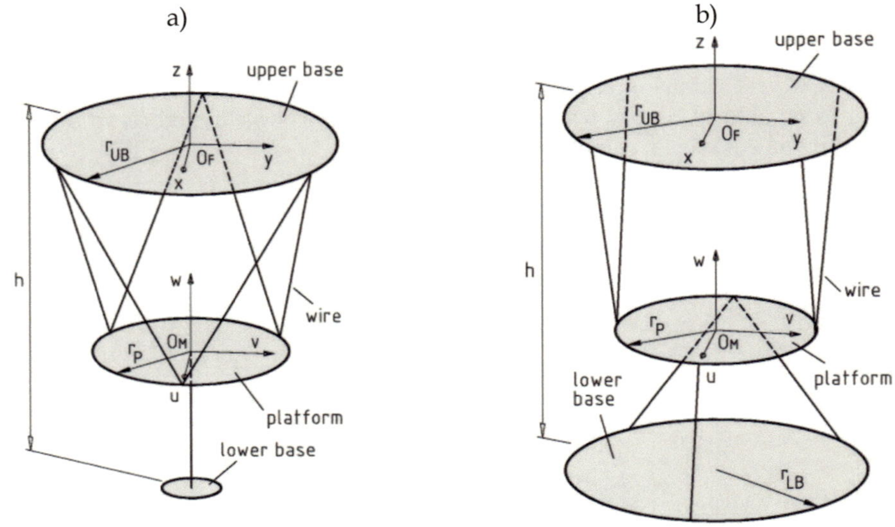

Fig. 2. Two examples of seven-cable parallel structures: a) WiRo-6.1; b) WiRo-4.3.

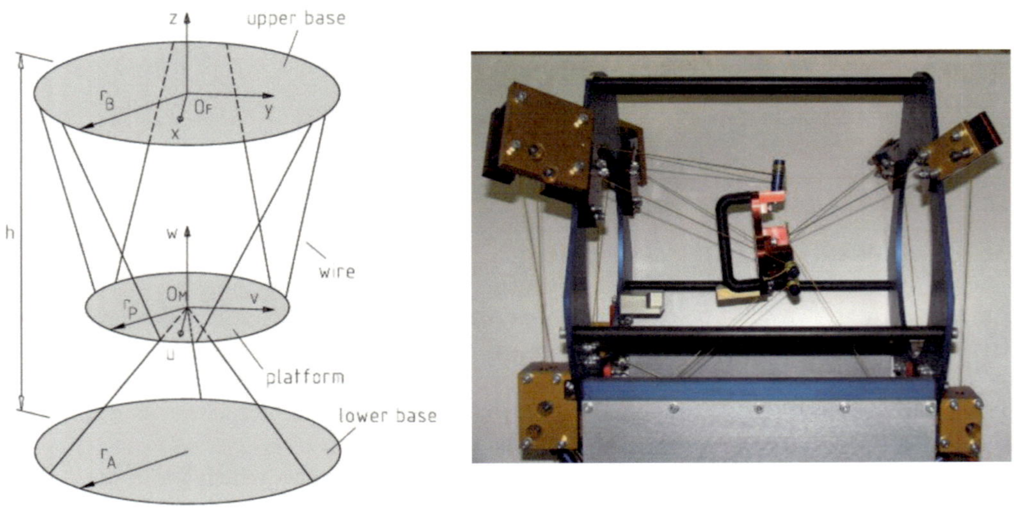

Fig. 3. A nine-cable parallel structure with polar symmetry and a prototype based on the same scheme: the WiRo-6.3.

Each of the structures is characterized by two coordinate systems, one integral with the fixed frame, with centre O_F and axes x, y, z, and the other moving with the platform, with centre O_M and axes u, v, w. The same notations apply to the nine-cable structure presented in figure 3, which has led to the realisation of the prototype shown beside. This device presents a similar layout to the one shown in figure 2a, with the single lower cable substituted by three cables converging to a single point on the platform. From the contraction of *Wire Robot* and from the layout of the cables (in number of p on the upper base, q on the lower one) the structures presented have been nicknamed WiRo-$p.q$ (Ferraresi et al., 2004).
The inverse kinematics study, providing the length of each actuator starting from the pose of the platform, is always simple for purely parallel structures. The following procedure does not only apply to the three structures shown as examples, but to any cable-driven robot and to any parallel device in general. It can be described through the simple geometric chain shown in figure 4, constituted by the fixed passing point P_{Fi}, the moving attachment point P_{Mi} and the cable linking them.

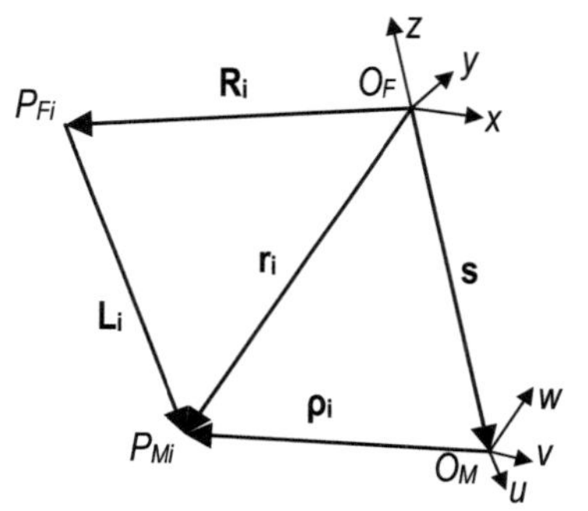

Fig. 4. Single kinematical chain of a parallel device.

Knowing the coordinates of P_{Fi} and P_{Mi} in their respective coordinate systems, the position vector representing the i-th mobile attachment point with respect to the fixed coordinate system is:

$$\mathbf{r_i} = \begin{bmatrix} x_i & y_i & z_i \end{bmatrix}^T = \mathbf{A} \cdot \boldsymbol{\rho}_\mathbf{i} + \mathbf{s} \tag{1}$$

where $\boldsymbol{\rho}_\mathbf{i} = \begin{bmatrix} u_i & v_i & w_i \end{bmatrix}^T$ is the position vector of the i-th attachment point with respect to the mobile coordinate system, $\mathbf{A}$ is the 3x3 orientation matrix of the platform and $\mathbf{s}$ is the position vector of the origin O_M. Naming $\mathbf{R_i} = \begin{bmatrix} X_i & Y_i & Z_i \end{bmatrix}^T$ the position vector of the passing point P_{Fi} with respect to the fixed frame, simple geometrical considerations lead to the vector representing the i-th cable:

$$\mathbf{L_i} = \mathbf{r_i} - \mathbf{R_i} \tag{2}$$

The modulus of $\mathbf{L_i}$ is the length of the i-th cable.

Contrary to the inverse kinematics, the forward kinematics - determining the pose of the platform from a given set of actuator lengths - is often quite complicated for parallel structures. In particular, it is not always possible to obtain a closed-form solution, obliging to work it out through numerical analysis. When designing the control software, this can become a huge issue since cycle times are critical in real-time applications. However, particular cases exist for which a closed-form solution can be found, depending on a convenient layout of the fixed and moving points.
For example, the nine-cable structure WiRo-6.3 presents a closed-form solution of the forward kinematics, thanks to the planarity of all moving attachment points and to the fact that three of them merge into one. This allows the three translation degrees of freedom of the platform to be decoupled from the orientation ones.
In the following a closed-form solution for the forward kinematics of the WiRo-6.3 is described (Ferraresi et al., 2004). The following approach does not require the polar symmetry of the robot; therefore, it can be used for any nine-actuator robot (not just cable robots) with six actuators connected to the same mobile platform plane and three actuators converging to a single point on the same plane.
As said above, the position of the centre of the platform O_M can be determined with ease. In fact, for each of the three lower cables it is $\boldsymbol{\rho}_\mathbf{i} = [0 \quad 0 \quad 0]^T$ (see figures 3 and 4). The equations that must be solved in order to obtain the vector $\mathbf{s} = [s_x \quad s_y \quad s_z]^T$ linking O_F to O_M are:

$$(\mathbf{s} - \mathbf{R}_\mathbf{i})^2 = \mathbf{s}^2 + \mathbf{R}_\mathbf{i}^{\,2} - 2\mathbf{R}_\mathbf{i} \cdot \mathbf{s} = L_i^{\,2} \tag{3}$$

For each of the three lower cables, the values of L_i and $\mathbf{R}_\mathbf{i}$ are different (but known), leading to a system of three equations in the form (3) with the three components of $\mathbf{s}$ as unknown quantities. Its solution is trivial and will not be exposed here for the sake of brevity.
To obtain the orientation matrix equations (1) and (2) are combined:

$$\mathbf{L}_\mathbf{i} = \mathbf{r}_\mathbf{i} - \mathbf{R}_\mathbf{i} = \mathbf{A} \cdot \rho_\mathrm{i} + \mathbf{s} - \mathbf{R}_\mathbf{i} = [L_{ix} \;\; L_{iy} \;\; L_{iz}\,]^T \tag{4}$$

Considering each component separately:

$$\begin{aligned} L_{ix} &= A_{11}\,u_i + A_{12}\,v_i + A_{13}\,w_i + s_x - X_i \\ L_{iy} &= A_{21}\,u_i + A_{22}\,v_i + A_{23}\,w_i + s_y - Y_i \\ L_{iz} &= A_{31}\,u_i + A_{32}\,v_i + A_{33}\,w_i + s_z - Z_i \end{aligned} \tag{5}$$

The length of the i-th cable is defined by the 2-norm of vector $\mathbf{L}_\mathbf{i}$. Squaring it:

$$\begin{gathered} L_i^2 = (\mathbf{L}_\mathrm{i})^\mathrm{T} \cdot (\,\mathbf{L}_\mathrm{i}) = L_{ix}^2 + L_{iy}^2 + L_{iz}^2 = \\ = s_x^2 + s_y^2 + s_z^2 - 2(s_x X_i + s_y Y_i + s_z Z_i) + r_P{}^2 + r_B{}^2 + 2(A_{11}\,u_i + A_{12}\,v_i + A_{13}\,w_i)(s_x - X_i) + \\ + 2(A_{21}\,u_i + A_{22}\,v_i + A_{23}\,w_i)(s_y - Y_i) + 2(A_{31}\,u_i + A_{32}\,v_i + A_{33}\,w_i)(s_z - Z_i) \end{gathered} \tag{6}$$

This formulation leads to a system of six equations, corresponding to each of the upper cables, in which the unknown quantities are the nine terms A_{ij}. In fact, all other quantities are known, and the three lower cables have already been used to find $\mathbf{s}$. However, three of the terms A_{ij} (A_{13} , A_{23} , A_{33}) disappear when considering the fact that $w_i = 0$ for every i, thanks to the planarity of the attachment points on the platform. A solution of the 6x6 system can now easily be found.

3. Workspaces

When designing a robot, particular care should be devoted to verify its operative capabilities, in particular its workspace and dexterity. In fact, a device that can work just in a very small portion of space, or with limited angles, is of little practical use. Furthermore, analysing cable-driven structures, it is not sufficient to consider the usual definition of workspace as *the evaluation of the position and orientation capabilities of the mobile platform with knowledge of the dimensional parameters, the range of actuated variables and the mechanical constraints*. In fact, a further limitation lays in the condition that cables can only exert traction forces. Thus the workspace of a cable-actuated device may be defined as the set of points in which the static equilibrium of the platform is guaranteed with positive tension in all cables (or tension greater than a minimum positive value), for any possible combination of external forces and torques.

At first, it can be supposed that cable tensions and external forces and torques can virtually reach unlimited values. Under that condition, the set of positions and orientations that the platform is able to reach can be called *theoretical workspace*.

To verify the possibility to generate any wrench with positive tensions in the cables, it is necessary to write the equations relating the six-dimensional wrench vector to the m-dimensional cable tension vector (with m: number of cables). The ability of any given device to provide a stable equilibrium to the end effector is called *force closure*. The force closure of a parallel structure in a particular configuration is calculated through the equation of statics:

$$-\mathbf{W} = \mathbf{f} = \tilde{\mathbf{J}} \cdot \boldsymbol{\tau} \tag{7}$$

where, in the case of a redundant parallel robot with m actuators, $\mathbf{W}$ is the six-component wrench acting on the platform, $\mathbf{f}$ is the wrench provided by the robot, $\tilde{\mathbf{J}}$ is the 6xm structure matrix calculated for any particular configuration and $\boldsymbol{\tau}$ is the m-component vector containing the forces of the actuators or, in the case of a cable-driven robot, the cable tensions.

The condition to check if a given pose of the platform belongs to the theoretical workspace can be expressed imposing that for any $\mathbf{f}$ the tensions of the cables can all be made positive (or greater than a prefixed positive value):

$$\boldsymbol{\tau} > 0 \tag{8}$$

while checking at the same time that $\tilde{\mathbf{J}}$ has full rank, equal to six (if not, the structure lays in a singular pose). Since $\tilde{\mathbf{J}}$ is not square, equation (7) allows an infinite number of solutions for any given $\mathbf{f}$. By inverting equation (7), the minimum-norm solution can be obtained:

$$\boldsymbol{\tau}_{\min} = \tilde{\mathbf{J}}^{+} \cdot \mathbf{f} \tag{9}$$

where $\tilde{\mathbf{J}}^{+}$ is the pseudoinverse of $\tilde{\mathbf{J}}$.
The generic solution of equation (7) is given by:

$$\boldsymbol{\tau} = \boldsymbol{\tau}_{min} + \boldsymbol{\tau}^* \tag{10}$$

where $\boldsymbol{\tau}^*$ must belong to the kernel, or null space of $\tilde{\mathbf{J}}$, defined through the expression:

$$\tilde{\mathbf{J}} \cdot \boldsymbol{\tau}^* = 0 \tag{11}$$

If $\tilde{\mathbf{J}}$ is not square, like in this case, the number of solutions of (7) is ∞^{m-6}. This means that the infinite possible values of $\boldsymbol{\tau}$ can be found by adding to $\boldsymbol{\tau}_{min}$ a vector $\boldsymbol{\tau}^*$ that does not affect the resulting wrench, but can conveniently change the actuator forces.
Condition (8) may be met for a particular six-dimensional point of the workspace if at least one strictly positive $\boldsymbol{\tau}^*$ exists. In this way, knowing that all the multiples of that $\boldsymbol{\tau}^*$ must also belong to the null space of $\tilde{\mathbf{J}}$, it is possible to find an appropriate positive multiplier c able to compensate any negative component of $\boldsymbol{\tau}_{min}$:

$$\mathbf{f} = \tilde{\mathbf{J}} \cdot (\boldsymbol{\tau}_{min} + c \cdot \boldsymbol{\tau}^*) \tag{12}$$

where, as said above:

$$c \cdot \boldsymbol{\tau}^* \in Null(\tilde{\mathbf{J}}); \quad \boldsymbol{\tau}_{min} + c \cdot \boldsymbol{\tau}^* > 0 \tag{13}$$

Having defined a convenient procedure to evaluate if a particular six-dimensional point belongs to the theoretical workspace, it is now possible to apply it to a discretised volume. It is not trivial to find out whether at least one strictly positive τ^* exists, especially for highly redundant structures; a possible method has been developed by the authors (Ferraresi et al., 2007) but its description is beyond the scopes of this Chapter and will not be presented here.
Moreover, several strategies may be adopted to minimise calculation times and to deal with displacements and orientations of the platform. In fact, since workspaces are six-dimensional sets it is not simple to represent them graphically. In order to obtain a convenient graphical representation, a possible choice is to consider separately the orientation and position degrees of freedom by distinguishing the *positional workspace* from the *α-orientation workspace*.
The positional workspace is the set of platform positions belonging to the workspace with the platform parallel to the bases. The α-orientation workspace is the set of platform positions that belong to the workspace for each of the possible platform rotations of an angle ±α around each of its three reference axes. With those definitions, both the positional and the α-orientation workspaces are three-dimensional sets.
As an example, figure 5 shows the positional workspace of the structures presented in figures 2a, 2b and 3, with their projections on the coordinate planes for visual convenience. Figure 6 shows their α-orientation workspaces for a few different values of α.

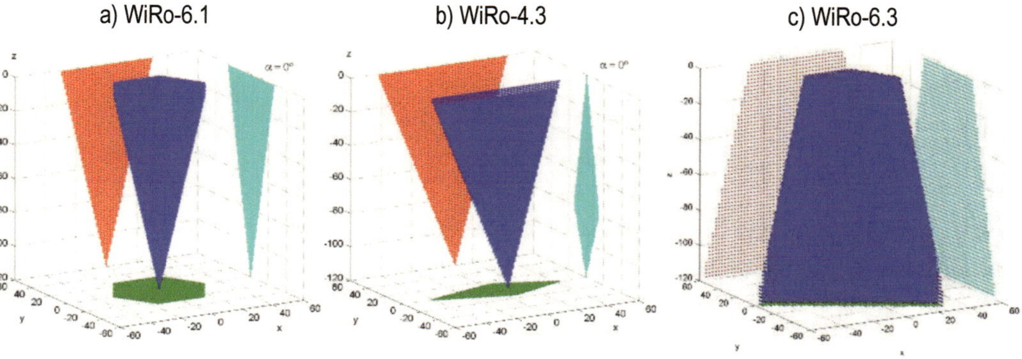

Fig. 5. Positional workspace of the three structures considered.

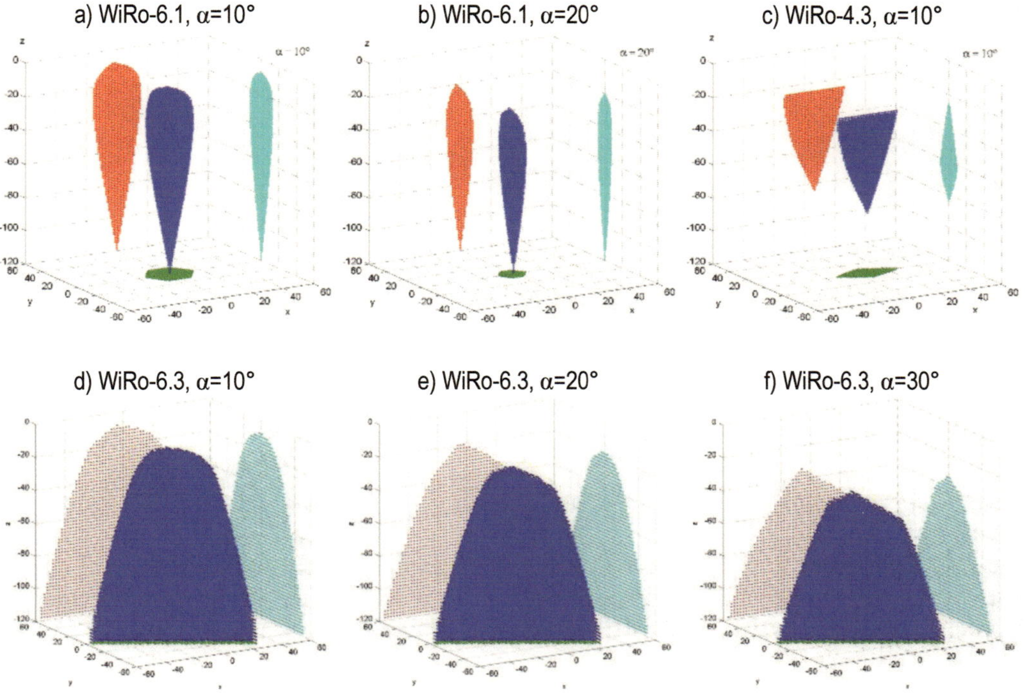

Fig. 6. α-orientation workspaces of the three structures for different values of α.

The geometric dimensions of the three structures have been set using arbitrary units, making them scalable. Obviously though, a rigorous method to compare the results is needed and it must be independent from the size and proportions of the structures.

Three dimensionless indexes have been proposed (Ferraresi et al., 2001) in order to analyse the results in a quantitative and objective way. They are the index of volume, the index of compactness and the index of anisotropy. The index of volume I_v evaluates the volume of the workspace relatively to the overall dimension of the robotic structure. The index of compactness I_c is the ratio of the workspace volume to the volume of the parallelepiped circumscribed to it. The index of anisotropy I_a evaluates the distortion of the workspace with

respect to the cube with edge equal to the average of the edges of the parallelepiped.
The mathematical expressions for those indexes are:

$$I_v = \frac{p\Delta x \Delta y \Delta z}{\pi h_{cc} D_{cc}^2 / 4} \qquad I_c = \frac{p\Delta x \Delta y \Delta z}{abc} \qquad I_a = \frac{|m-a| + |m-b| + |m-c|}{m} \tag{14}$$

where p is the quantity of discrete points contained into the workspace, Δx, Δy, Δz are the discretisation steps used along their respective axes, D_{cc} and h_{cc} are the base diameter and height of the smallest cylinder containing the whole structure, a, b, c are the edges of the parallelepiped circumscribed to the workspace, and m is the average of a, b and c.
An optimal workspace should have large indexes of volume and compactness, and an index of anisotropy as close as possible to zero. As an example, these three indexes can be used to compare the workspaces of the three devices considered above, shown in figures 5 and 6.

Structure	α	I_v	I_c	I_a
WiRo-6.1	0°	0.07	0.26	1.1
	10°	0.02	0.35	2
	20°	0.006	0.34	2.5
	30°	0.0004	0.28	3.2
WiRo-4.3	0°	0.04	0.18	1.3
	10°	0.007	0.23	1
	20°	0	0	NaN
WiRo-6.3	0°	0.31	0.4	0.17
	10°	0.24	0.34	0.24
	20°	0.18	0.3	0.36
	30°	0.06	0.3	0.85

Table 1. Application of volume, compactness and anisotropy indexes to the three structures.

Comparing figures 5 and 6, the different performance of the structures in terms of workspaces is evident. Table 1, thanks to the three indexes, provides a more rigorous support for the comparative evaluation of different devices.

4. Force reflection

Any cable-driven structure of the kind presented in Section 2 may be used as an active robot, installing an end effector on the platform and controlling its pose through the imposition of cable lengths. However, on the contrary, it may also work as a master device for teleoperation: for this, a handle or similar object must be integrated on the platform to allow command by an operator. In this case the user determines the pose of the platform which in turn constrains the theoretical cable lengths.
To avoid any cable to be loose, all of them must be continuously provided with a pulling force greater than zero; moreover, it is not enough to provide a constant tensioning force to

each cable because, due to their different orientations, the resulting wrench on the platform might greatly disturb the user's operation.
So, apart from peculiar cases of little interest here, every cable must be actuated by winding it to a spool integral to a rotary motor shaft, or directly attached to a linear motor or any other convenient actuation source.
Since the aim is controlling the resultant wrench on the platform, each actuator pulling a cable must be force- or torque-controlled (opposed to the case of an active robot, where the control imposes positions and velocities and forces and torques come as a consequence). Through a convenient set of cable tensions it is possible to impose any desired wrench on the platform and, finally, on the user's hand. The first, intuitive choice could be setting to zero all forces and torques acting on the platform, to permit the user an unhampered freedom of movement. However, it is more interesting to provide the device with force reflection capabilities.
The presence of force reflection in a teleoperation device gives the operator a direct feeling (possibly scaled) of the task being performed by the slave device. In this way, effectiveness of operation improves greatly, because the operator can react more promptly to the stimuli received through the sense of touch than if he had only visual information, even if plentiful (direct eye contact, displays, led indicators, alarms, etc.). For example, it is not immediate to perceive the excessive weight of a remotely manipulated object, or a contact force unexpectedly high, using only indirect information; when the alarm buzzes, or the display starts flashing, it might already be too late. On the contrary, if forces and torques are directly reflected to the operator, he might act before reaching critical situations. The same applies for small-scale teleoperation, e.g. remote surgery: excessive forces may have terrible consequences.
Equations (12) and (13) guarantee that it is theoretically possible to give the platform any desired wrench, if its current pose belongs to the theoretical workspace.
Statics relates the cable forces to the six-dimensional wrench on the platform, according to equation (7). For a nine-cable structure it can be interpreted as follows: given a vector $\mathbf{f} \in R^6$ that is desired to act on the platform as a force reflection, it is necessary to find a vector of cable forces $\boldsymbol{\tau} \in R^9$ fulfilling equation (7). Due to the redundancy of the structure, if $\tilde{\mathbf{J}} \in R^{6x9}$ has a full rank equal to 6, the set of vector fulfilling equation (7) is a three-dimensional hypersurface in a nine-dimensional Euclidean space, meaning that the number of solutions is ∞^3.
Among all possible solutions, the one reckoned optimal may be chosen through the following considerations. Once a minimum admissible cable tension τ_{adm} has been set, every component of $\boldsymbol{\tau}$ must be greater than or equal to that value, while at the same time keeping them as low as possible and still fulfilling equation (7).
Therefore the following target may be written:

$$\text{minimize } G = \sum_{i=1}^{9} \tau_i \tag{15}$$

under the conditions:

$$\begin{cases} \tilde{\mathbf{J}} \cdot \boldsymbol{\tau} = \mathbf{f} \\ \tau_i \geq \tau_{adm} \quad i = 1...9 \end{cases} \tag{16}$$

That is a linear programming problem that may be solved, for instance, by using the simplex method. The solution to the problem (15), (16) leads to an optimised and internally connected $\boldsymbol{\tau}$, i.e. it can be demonstrated that if $\mathbf{f}$ and $\tilde{\mathbf{J}}$ vary continuously, then also the solution $\boldsymbol{\tau}$ calculated instant by instant presents a continuous run against time.

The procedure to identify the theoretical workspace does not take into account the interaction of the structure with the environment, in terms of maximum forces and torques acting on the platform, and the maximum tension each cable can exert. Therefore a further, different definition of workspace is necessary, involving those considerations. The portion of theoretical workspace where the structure can provide the desired wrench with acceptable cable tensions is called *effective workspace*.

In detail, to find that out, the following parameters must be set: maximum force on the operator's hand in any direction, maximum torque around any axis, minimum and maximum admissible values of cable tension. Then, for every pose in the theoretical workspace, maximum forces and torques must be applied in different directions. For every pose, the cable tensions must be calculated according to the problem (15), (16), recording the largest value of cable tension. In this way, every pose of the platform is characterised by a maximum cable tension resulting from the application of the maximum wrench. This value can be compared to the maximum admissible one, determining whether or not that particular pose belongs to the effective workspace.

As an example, figure 7 shows in graphical form a few results created applying that procedure to the WiRo-6.3, for a given set of parameters (maximum force on the operator's hand in any direction: 10N, maximum torque around any axis: 1Nm, minimum admissible value of cable tension: 5N, maximum value of cable tension: 150N). For the sake of graphical representation, the workspace has been cross sectioned at various values of z; the base plane represents the platform centre position on that cross section, while the dimension on the third axis and the colour intensity represent the cable tension magnitude.

After a complete scan of the workspace, the result is – in this particular case – that the effective workspace is a wide subset of the theoretical one, making it possible to construct a structure with the physical characteristics that have been chosen as parameters. On the other hand, it must be noted that towards the borders of the workspace the maximum tensions increase dramatically, resulting one or even two orders of magnitude greater than in the central portion. Therefore, possible misuse of the structure taking the platform in one of those conditions must be carefully avoided; otherwise cable tensions and forces on the operator's hand can literally become uncontrollable. Obviously, the same should be done for orientations which, in the examples considered here, must be limited to ±30° around any axis (a greater angle would dramatically reduce the available orientation workspace shown in figure 6). A possible strategy can be generating a strong opposing force (or torque) when the operator tries to move (or rotate) the platform across the border of the effective workspace, thus limiting its freedom of movement "virtually", i.e. without the use of physical end-of-run stop devices.

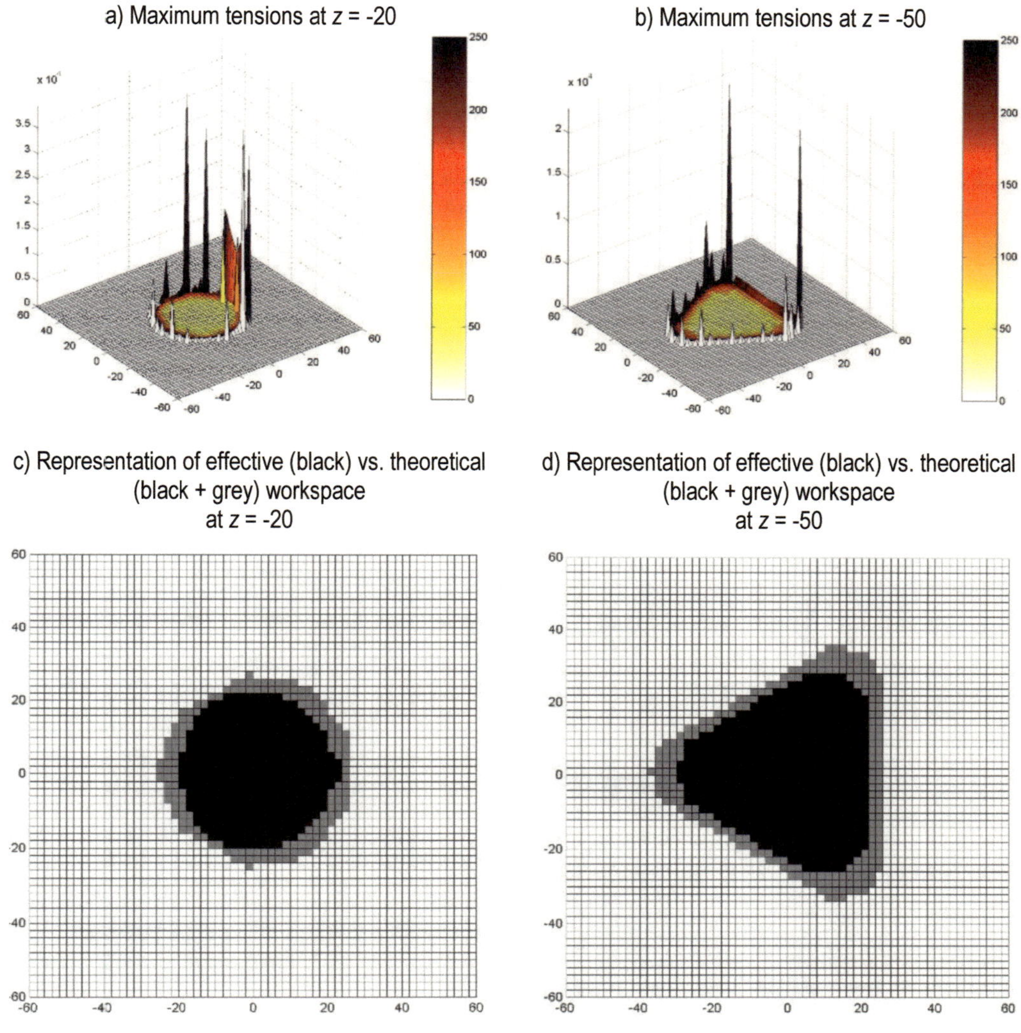

Fig. 7. a), b) Example cross sections of the workspace showing maximum cable tensions. c), d) The same cross sections shown to underline the distinction between theoretical and effective workspace.

5. Device control and cable actuation

The control logic is summarized in figure 8. The operator imposes position and velocity to a proper element, which may be a handle or some other device, suspended in space by the cables. Each cable is tensioned by a specific actuator and under the operator's action it can vary its length between the fixed and moving points (indicated as P_{Fi} and P_{Mi} in figure 1).
Measuring the length of each cable through transducers, the control system is able to evaluate position, orientation, linear and angular velocity of the handle by means of the forward kinematics algorithm. Those results are used as reference input to the control

system of the slave robot actuators: the slave unit is therefore driven to carry out a particular task.

As a consequence, interaction forces are exerted on the slave unit by the environment; such forces may be measured by convenient sensors and reproduced on the operator's handle. The global environment forces are therefore used as reference input to the control system. The latter can calculate the exact force that each cable must exert on the handle by means of the inverse statics algorithm; this value is therefore used as reference input to each cable actuator.

The control system may apply a scale factor both to the movement of the slave unit and to the forces reflected to the operator.

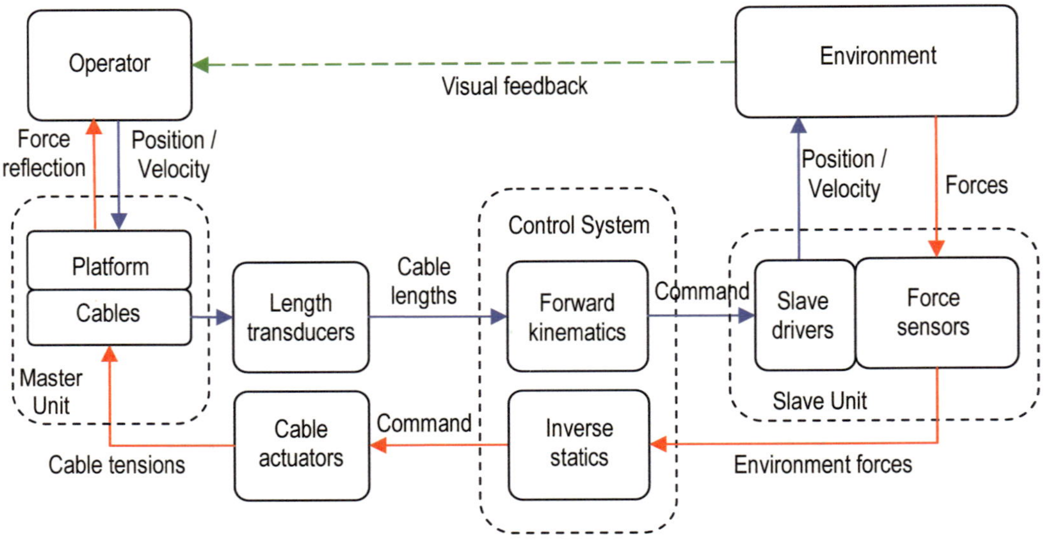

Fig. 8. The control logic of a telemanipulation system with a cable driven haptic master

The control system must perform a double function:

(a) on the basis of the cable lengths it must calculate the pose of the platform; since the master has a parallel structure (that may be also multi-redundant), this real-time calculation can be very demanding;

(b) a convenient sensorial system should read the wrench exerted by the environment on the remote slave unit; such six-degree-of-freedom information is used by the control system to calculate, by means of an inverse statics algorithm, the exact value of each cable tension; again, the single or multiple redundancy of the parallel cable structure makes real-time operation particularly demanding.

The accuracy of the control depends on the physical characteristics of the device, in particular the mechanical layout and how the control system interacts with the cables and their actuators.

The two main functions of the control system (master pose calculation and cable tension generation) may be affected by certain errors, which can compromise the effectiveness of the device.

In particular, the kinematical accuracy is strictly related with the characteristics of the cables and their path. The path of each cable can be described considering two different sections, starting from the mobile platform where one end is attached. The first is called the *free section* of the cable, since its direction and its length are determined by the platform motion. The cable passes through a given point on the structure corresponding theoretically to a fixed point. The variable lengths of the free section of each cable are used to calculate the platform pose. The second section of the cable goes from the cited fixed point to the actuation group and has an ideally constant length.
The way to realize the passage through the fixed point is crucial for the geometric accuracy of the system.
Figure 9 shows a possible way to realize a fixed passing point of the cable. The cable is conveyed into a cylindrical nylon insert whose entrance hole represents the theoretical passing point and then it is diverted by the pulley towards the actuator.

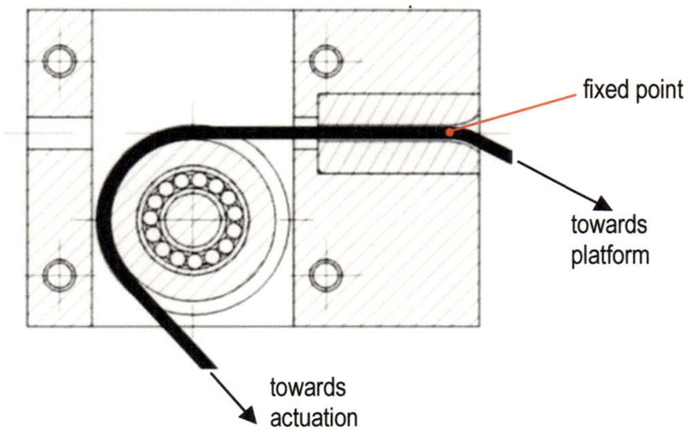

Fig. 9. Realization of a fixed passing point for the cable using a low-friction insert

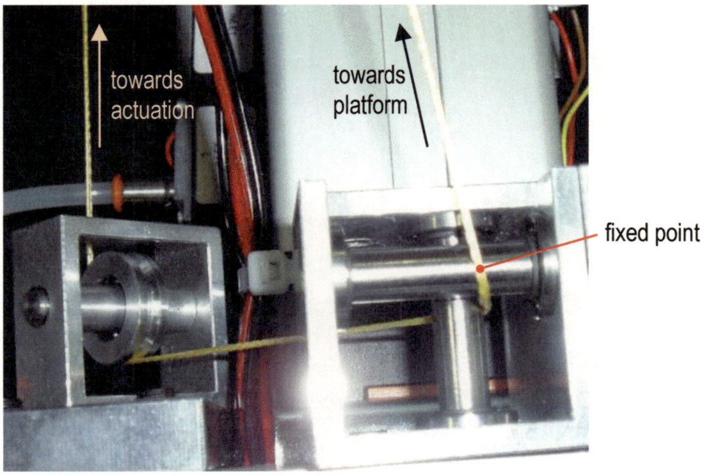

Fig. 10. Realisation of a fixed passing point for the cable using skew rollers.

Another possible solution is shown in figure 10. The cable coming from the platform is diverted by a pair of skew rollers mounted perpendicularly, and then follows the second part of the path defined by fixed pulleys.

Due to the physical characteristics of cables and diversion system, during the handle motion the end of the cable free section is not actually fixed, but moves in a given range, introducing an error in the calculation of the platform pose. For instance, the solution of figure 10 is less accurate than the former one but greatly reduces friction, producing a considerable advantage as will be explained later.

In an ideal situation, each cable should be inextensible and perfectly flexible; as a matter of fact, tensile load causes deformation and flexural stiffness makes sure that the theoretical passing point actually corresponds to a small area. To limit such drawbacks it is necessary to choose cables with convenient characteristics. In particular, a good choice may be adoption of synthetic fibre cables, such as Dyneema® (Hoppe, 1997). In comparison with steel wires of the same strength, those cables are much more flexible and just slightly more extensible, while presenting a similar cross section.

The greater flexibility reduces drastically the extension of the passing area, which can reasonably be approximated to a point.

Moreover, the longitudinal compliance of Dyneema cables is comparable with steel cables but, although non-linear, presents a much more regular run, thus allowing compensating the length variation with a convenient relation as a function of cable tension.

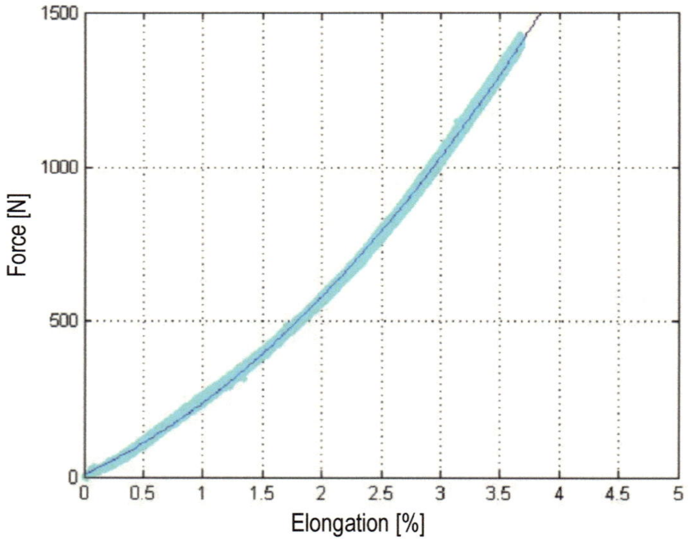

Fig. 11. Mechanical characteristic of a Dyneema-140 cable

As an example, figure 11 shows the mechanical characteristic of a Dyneema-140 cable. This characteristic allows calculating a simple function yielding the percentage elongation (E) as a function of the tensile force (F) expressed in N:

$$E = -1.53 + \sqrt{2.34 + 0.0177 \cdot F} \tag{17}$$

The non-linearity of the cable characteristic can also be exploited as an advantage. As can be seen in figure 11, such a stiffening run allows to make the whole structure more rigid by properly increasing the minimum tension τ_{adm} in the conditions (16), thus producing an advantage for the system controllability.
A further advantage of great cable flexibility is the simplicity of realising the attachment to the platform: this can be realized by a simple knot, practically reducing the attach area to a virtual point fixed to the handle.
As regards the fixed-length part of the cable, it covers a path from the fixed point to the actuator. It is important that the cable constraints are as rigid as possible; therefore the cable should preferably be guided by pulleys rather than a flexible sheath which may be longitudinally compliant.
Another peculiarity of haptic teleoperation systems is the fact that the accuracy of the master pose calculation is less important than that required for the force reflection on the operator's hand.
This is in analogy with actions effected directly by a human subject: the approach and positioning of the hand is controlled by the human proprioceptive (i.e. self-sensing, detecting the motion or position of the body or a limb) and visual system, relatively rough, while the completion of the operation, however accurate, is controlled by tactile forces.
The generation of the wrench on the operator's hand is therefore a very critical point, which requires great accuracy and must compensate various disturbances.
Friction must be avoided or limited as much as possible. Effective software compensation by means of an algorithm that considers the sliding direction and velocity of seven or more cables is practically impossible due to irregular behaviour and most of all to the discontinuity around zero velocity. That is a further reason in favour of realising the cable path along the fixed-length part using pulleys and low friction bearings rather than a flexible sheath.
Friction may be present not only in the cable path, but also inside the actuators. Therefore their choice is a very strategic point. Generally, two main options are available: (a) electric motors; (b) pneumatic actuators.
The most common solution in this kind of application is the adoption of electric motors, but this choice is somehow disputable: the actuators must operate around the stall condition, so they may have difficulties in controlling accurately the torque value.
A typical drawback arises when using brushless motors, characterized by a more or less evident *cogging torque*, i.e. a variation in torque depending on the rotor angular position. In some cases such variation can amount to 10% of the total torque.
Moreover, a rotating motor requires a given device to transform its motion into linear, for example a spool on the shaft. On the other hand, an electric motor presents a very high dynamics, thus allowing an effective control of the cable tension during the handle motion.
An effective alternative is represented by pneumatic actuators. This technology may give interesting advantages, but several aspects need to be accurately considered and evaluated (Ferraresi et al., 2006)
Force control is apparently simpler, since this is obtained by controlling the air pressure in the chambers of a cylinder. Such operation is relatively easy in static condition, but could be difficult during motion, because of the low dynamics of the overall pneumatic system (actuator, valves, piping).

Pneumatic actuators present noticeable friction, due to the seals and also to the fluidic resistance of the air through the various orifices; even the adoption of a membrane pneumatic actuator cannot completely eliminate friction.
The motion of pneumatic actuators is already linear, but may be too limited with respect to the requirements; in this case the actuator stroke can be multiplied by means of a pulley device of the kind shown in figure 12. In this device, the actuator 1 exerts his action on a group of mobile pulleys 2; the cable 3 leans on mobile pulleys 2 and fixed pulleys 4-5; a rotating potentiometer 6, connected to the fixed pulley 5, measures the cable motion. In this example the actuator stroke is multiplied by 6, and its force is divided by the same factor. Such a device allows limiting the size of the actuator, but on the other side it introduces further friction in the system. Of course the stroke multiplier may also be used when adopting linear electric motors, in order to limit their size.

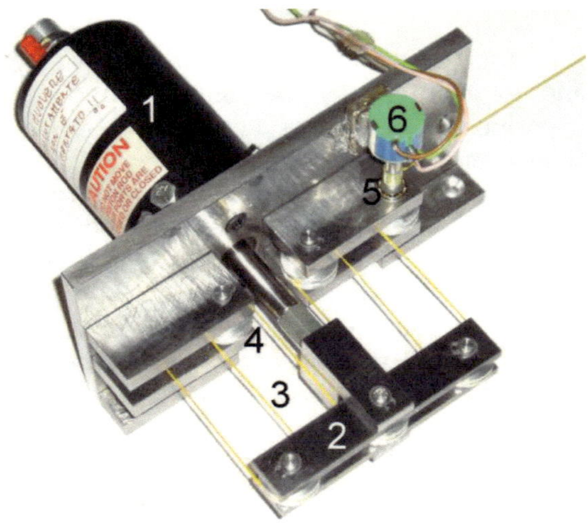

Fig. 12. Device for stroke multiplication of a linear actuator.

In case of pneumatic actuation of the cables, particular attention is required for the control of the supply pressure of the actuator. A simple and economic way is to use on-off valves controlled by PWM logic, but it is necessary to achieve a good compromise, considering the following points:

(a) the PWM signal introduces vibrations in the pressure value, generating the force reflection; such vibrations may be reduced by a proper dimensioning of the pneumatic system, creating a low-pass pneumatic filter that, on the other side, may penalize the system dynamics;
(b) the overall dynamics also depends on the valve size; in particular the discharge valves should be properly dimensioned.

As regards the control of cable tensions, it is possible to choose two different strategies:

(a) Open loop control. In this case the tension value calculated by the control system is used as reference for the motor torque (in case of electrical actuation of cables) or the cylinder pressure (in case of pneumatic actuation). The actual cable tension will be therefore affected by errors due to the friction along the transmission line and various disturbances such as cogging torque, PWM control, etc.

(b) Closed loop control. The reference value is compared to a direct measure of the cable tension, through a convenient sensor. The control is in general more accurate, but instabilities may arise, due to the non-linearities present in the system, such as friction, cogging torque of electric motors, static and dynamic characteristics of pneumatic valves.

Among all the possible options presented in the previous paragraphs, a choice must be made according to the project specifications, desired performances and, of course, budget limitations.

6. Conclusion

Cable driven robotic devices present peculiar aspects, requiring the solution of very specific problems.

As regards the determination of the main operational characteristics, like the workspace, it is necessary to consider that such devices are often parallel and redundant structures. Moreover, cables can only exert traction forces. As a consequence, the development of such structures requires coping with two main aspects:

(a) the solution of forward kinematics and inverse statics, necessary for the remote control of the slave device and the force reflection on the operator, are usually very complicated and often a closed form does not exist;

(b) it is useful to find a convenient layout of the attach points on the mobile platform, in order to obtain closed-form solutions or optimal numerical solutions for the kinematical and static equations.

The workspace characteristics also depend on the number of cables and the layout of their passing points on the fixed structure, but it is difficult to foresee those performances in the design phase and to evaluate them in an objective way. Therefore it is necessary to develop methods aimed at finding out the shape of the six-dimensional workspace, expressing it in a graphical form and evaluating its characteristics by means of objective numerical parameters.

The actuation by cables needs particular attention as regards both the generation and the transmission of cable tensions. This has required a wide research activity concerning the kind of actuators to be used and the mechanical aspects of the transmission lines.

As regards the actuators, it must be noticed that their action must be generated at very low velocities, often near the stall condition. Therefore, particular care must be devoted to the choice of technology (electrical or pneumatic), to the kind of actuator and to the way of controlling the force or torque.

The physical characteristics of the cable and its path have direct influence on the control accuracy. The cables connect the operator with the motors and with the device sensors; the accuracy of the device is affected by the friction acting on the cable and by characteristics like longitudinal compliance and flexural stiffness; those phenomena may only be partially compensated by the control system, therefore the choice of the cable is a very delicate point.

The research described here has faced all the aspects above mentioned. Several original solutions reported have been described in detail in previous works.

Future work will be devoted to improve the way of controlling the reflection forces, which has been spotted as the most delicate aspect for the accuracy of this kind of devices.

Acknowledgements

The work presented here has been partially funded by the Robotics and Telescience Research projects of the Italian Antarctic Research Program (PNRA).
A further support was provided by the Italian Ministry of University and Research.

7. References

Batsomboon, P.; Tosunoglu, S. & Repperger, D.W. (2000). A Survey of Telesensation and Teleoperation Technology with Virtual Reality and Force Reflection Capabilities, *International Journal of Modelling and Simulation*, Vol. 20, No. 1, (2000) pp. 79-88, ISSN 0228-6203

Bostelman, R.; Albus, J.; Dagalakis, N. & Jacoff, A. (1996). RoboCrane project: An advanced concept for large scale manufacturing, *Proceedings of AUVSI Conference*, Orlando, FL, July 1996

Conklin, W. & Tosunoglu, S. (1996). Conceptual Design of a Universal Bilateral Manual Controller, *Proceedings of 1996 Florida Conference on Recent Advances in Robotics*, pp. 187-191, Florida Atlantic University, Boca Raton, Florida, April 1996

Ferraresi, C.; Pastorelli, S. & Pescarmona F. (2001). Workspace analysis and design criteria of 6 d.o.f. wire parallel structures, *Proceedings of 10th International Workshop on Robotics in Alpe-Adria-Danube Region RAAD '01*, CD-Proceedings, paper RD-062, Vienna, Austria, May 2001

Ferraresi, C.; Paoloni, M.; Pastorelli, S. & Pescarmona, F. (2004). A new 6-DOF parallel robotic structure actuated by wires: The WiRo-6.3, *Journal of Robotics Systems*, Vol. 21, No. 11 (November 2004) pp. 581-595, ISSN 0741-2223

Ferraresi, C.; Carello, M.; Pescarmona, F. & Grassi, R. (2006). Wire-driven pneumatic actuation of a new 6-dof haptic master, *Proceedings of ESDA2006 8th Biennial ASME Conference on Engineering Systems Design and Analysis*, ISBN 0-7918-3779-3, Torino, Italy, July 2006

Ferraresi, C.; Paoloni, M. & Pescarmona, F. (2007). A new methodology for the determination of the workspace of six-DOF redundant parallel structures actuated by nine wires, *Robotica,* Vol. 25, No.1 (January 2007) pp. 113-120, ISSN 0263-5747

Hoppe, L.F.E. (1997). Performance improvement of Dyneema® in ropes, *Proceedings of OCEANS '97 MTS/IEEE Conference*, pp. 314-318, ISBN 0-7803-4108-2, Halifax, NS, Canada, October 1997

Kawamura, S.; Choe, W.; Tanaka, S. & Pandian, S.R. (1995). Development of an Ultrahigh Speed Robot FALCON Using Wire Drive System, *Proceedings of IEEE Int. Conference on Robotics and Automation*, pp. 215-220, ISBN 0-7803-1965-6, Nagoya, Aichi, Japan, May 1995

McAffee, D.A. & Fiorini, P. (1991). Hand Controller Design Requirements and Performance Issues in Telerobotics, *Proceedings of Advanced Robotics, 1991. Proceedings of Robots in Unstructured Environments, 91 ICAR*, pp. 186-192, ISBN 0-7803-0078-5, Pisa, Italy, June 1991

11

An original approach for a better remote control of an assistive robot

Sébastien Delarue, Paul Nadrag, Antonio Andriatrimoson, Etienne Colle and Philippe Hoppenot

IBISC Laboratory - University of Evry Val d'Essonne - France

Abstract

Many researches have been done in the field of assistive robotics in the last few years. The first application field was helping with the disabled people's assistance. Different works have been performed on robotic arms in three kinds of situations. In the first case, static arm, the arm was principally dedicated to office tasks like telephone, fax... Several autonomous modes exist which need to know the precise position of objects. In the second configuration, the arm is mounted on a wheelchair. It follows the person who can employ it in more use cases. But if the person must stay in her/his bed, the arm is no more useful. In a third configuration, the arm is mounted on a separate platform. This configuration allows the largest number of use cases but also poses more difficulties for piloting the robot.

The second application field of assistive robotics deals with the assistance at home of people losing their autonomy, for example a person with cognitive impairment. In this case, the assistance deals with two main points: security and cognitive stimulation. In order to ensure the safety of the person at home, different kinds of sensors can be used to detect alarming situations (falls, low cardiac pulse rate...). For assisting a distant operator in alarm detection, the idea is to give him the possibility to have complementary information from a mobile robot about the person's activity at home and to be in contact with the person. Cognitive stimulation is one of the therapeutic means used to maintain as long as possible the maximum of the cognitive capacities of the person. In this case, the robot can be used to bring to the person cognitive stimulation exercises and stimulate the person to perform them.

To perform these tasks, it is very difficult to have a totally autonomous robot. In the case of disabled people assistance, it is even not the will of the persons who want to act by themselves. The idea is to develop a semi-autonomous robot that a remote operator can manually pilot with some driving assistances. This is a realistic and somehow desired solution. To achieve that, several scientific problems have to be studied. The first one is human-machine-cooperation. How a remote human operator can control a robot to perform a desired task? One of the key points is to permit the user to understand clearly the way the robot works. Our original approach is to analyse this understanding through appropriation concept introduced by Piaget in 1936. As the robot must have capacities of perception decision and action, the second scientific point to address is the robot capacities of autonomy (obstacle avoidance, localisation, path planning…). These two points lead to propose different control modes of the robot by a remote operator, from a totally manual mode to a totally autonomous mode. The most interesting modes are the shared control modes in which the control of the degrees of freedom is shared between the human operator and the robot. The third point is to deal with delay. Indeed, the distance between the remote operator and the robot induces communication delays that must be taken into account in terms of feedback information to the user. We will conclude this study with several evaluations to validate our approach.

1. Introduction

Many researches have been done in the field of assistive robotics in the last few years. The first application field was helping with the disabled people's assistance. Different works have been performed on robotic arms in three kinds of situation. In the first case, static arm, the arm was principally dedicated to office tasks like telephone, fax… Several autonomous modes exist which need to know the precise position of objects. In the second configuration, the arm is mounted on a wheelchair. It follows the person who can employ it in more use cases. But if the person must stay in her/his bed, the arm is no more useful. In a third configuration, the arm is mounted on a separate platform. This configuration allows the largest number of use cases but also poses more difficulties for piloting the robot.

The second application field of assistive robotics deals with the assistance at home of people losing their autonomy, for example a person with cognitive impairment. In this case, the assistance deals with two main points: security and cognitive stimulation. In order to ensure the safety of the person at home, different kinds of sensors can be used to detect alarming situations (falls, low cardiac pulse rate…). For assisting a distant operator in alarm detection, the idea is to give him the possibility to have complementary information from a mobile robot about the person's activity at home and to be in contact with the person. Cognitive stimulation is one of the therapeutic means used to maintain as long as possible the maximum of the cognitive capacities of the person. In this case, the robot can be used to bring to the person cognitive stimulation exercises and stimulate the person to perform them.

Different works deal with autonomous robotics. They have several drawbacks in these kinds of application. Concerning disabled people assistance, persons want to act by herself/himself on the environment. In the case of people loosing their autonomy, one important point is to permit human-human interaction, through a robot seen as intermediary communication. One can also notice that autonomous robotics can not yet propose robots with several days of autonomy (except for spatial missions but at very expensive costs and with limited action capabilities). Our option is to develop remote control robots. That has two main advantages. As human being is in the control loop, it is possible to use her/his capacities, especially decision ones, which are the most difficult to get from a robot. That permits to assure total autonomy. The second point is that the remote operator is involved in the performed task, which is for example a clear demand from disabled people using technical assistance. This choice implies that a mission is realised by close Human Machine Cooperation between the robot and the human operator. It is also clear that the robot has some kinds of autonomous capacities, which can be used by the remote operator if needed.

The first part of this paper deals with autonomy capacities of the robot. It is essential to know what the robot can do to think about Human Machine Cooperation, which is the main point of the second part. We also propose different evaluations results in the last part of the paper.

2. Robot capacity of autonomy

Displacement in an environment is realised in two steps. The first one is the description of the trajectory to perform to reach the given goal. The second one consists in following the previous trajectory, avoiding unexpected obstacles. To make these two steps possible, the system needs to have information on its environment and the capacity to localise itself in this environment. We do not address the case in which the system has no information on its environment and builds itself a representation of it, using SLAM techniques. We suppose the system has a sufficient precise knowledge of the environment, which is the case in our application field. In short, displacement in an environment requires tow kinds of capacities: path planning and obstacle avoidance. A combination of these two capacities gives the robot navigation capacities. Localisation is also needed to achieve displacements toward a goal.

2.1 Trajectory planning

This is the first step of autonomy and permits to define the trajectory the robot has to follow. [Latombe91] presents the three main method families for planning: road maps, cell decomposition and potential fields. Before presenting them briefly, it is useful to define what free space is.

Free space describes all the positions the robot can reach taking into account environment information. In 2D mobile robotics, that represents all the (x, y, θ) positions where the robot can arrive. To simplify computation algorithms, a classical solution is to describe a kind of "growing" obstacle obtained by extension of the workspace by the dimensions of the robot. In that case, each orientation of the robot generates a workspace in which the robot can be considered as a point. In the case of a circular robot, only one workspace is needed. We introduced imaginary obstacles around the door to make the planned trajectory easier to follow.

The study of the connectivity of the robot's free space enable to determine a network of 1 dimension curves called a roadmap, which describe all possible trajectories from an initial point to a goal point. Among all possible paths, only one is chosen. A^* algorithm is well adapted for this work. Given a cost function, it determines the optimal path. It is possible to optimise the distance, but the function can also take into account the amount of energy, perception, capacities of the robot. In the case of cell decomposition, the workspace of the robot is split into several parts called cells. They are built to assure that all couples of points inside the same cell are linkable by a straight line. A graph linking all the adjacent cells is also built, which is called the connectivity graph. The idea of potential fields is to determine an artificial potential field representing the constraints given by the environment. Obstacles create a repulsive force while the goal creates an attractive force. This method has a well-known major drawback: the function has local minima, which are not the goal.

In our project, we use a visibility graph with the A* algorithm. In our application (the robot evolves indoors), the environment is sufficiently well-known so that the built graph is nearly complete. Moreover, cost function is interesting to use because we can choose several parameters to optimise the trajectory. All these aspects are detailed in [Hoppenot96] and [Benreguieg97].

2.2 Obstacle avoidance

A robot can follow the trajectory planned as above only if the environment is totally known. That is why local navigation systems have been developed based on robot sensors acquisition. It is now usual to combine path planning and local navigation. In that case, path planning is in charge of long terms goal while local navigation only deals with obstacle avoidance. Some qualitative reasoning theories have been developed. We propose a solution based on fuzzy logic to process obstacle avoidance ([Zadeh65], [Kanal88], [Lee90]).

When the vehicle is moving towards the target and the front sensors detect an obstacle located on the path, an avoiding strategy is necessary. The selected method consists in reaching the middle of a collision-free space. The used navigator is built with a fuzzy controller based on a set of rules as follows:

If Rn is x_i **and** Ln is y_i **Then** ω is t_i

and if Fn is z_i **then**

v is u_i "

else...

x_i, y_i, z_i, t_i and u_i are a linguistic labels of a fuzzy partition of the universes of discourse of the inputs Rn, Ln and Fn and the outputs ω and v, respectively. The inputs variables are the normalised measured distances on the right R, on the left L and in front F such as:

$$R_n = \frac{R}{R+L},\ L_n = \frac{L}{R+L} \text{ and } F_n = \frac{F}{\text{inf}}$$

where inf is the sensor maximum range.

The output variables are the angular and the linear speeds. On simplicity grounds, the shape of the membership functions is triangular and the sum of the membership degrees for each variable is always equal to 1. The universes of discourse are normalised between -1 and 1, for ω, and 0 and 1 for the other ones.

Each universe of discourse is shared in five fuzzy subsets. The linguistic labels are defined as follows:

Z	: Zero	NB	: Negative Big
S	: Small	NS	: Negative Small
M	: Medium	Z	: Zero
B	: Big	PS	: Positive Small
L	: Large	PB	: Positive Big

The whole control rules deduced from a human driver's intuitive experience is represented by fifty rules shown in the two following decision tables (Table1 and Table2): 25 rules allow to determine the angular velocity w and 25 others determine the linear speed v.

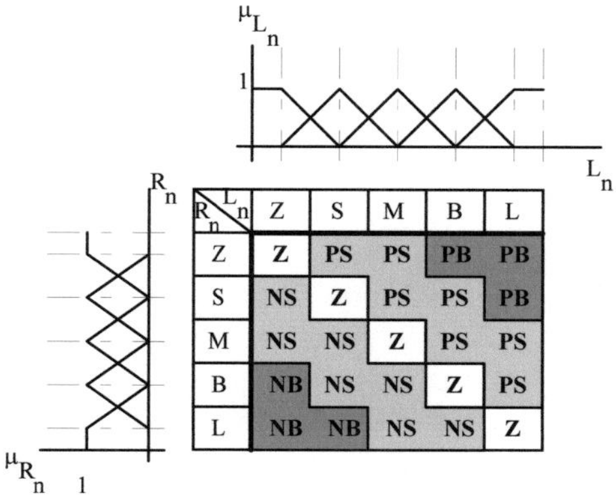

R_n \ L_n	Z	S	M	B	L
Z	Z	PS	PS	PB	PB
S	NS	Z	PS	PS	PB
M	NS	NS	Z	PS	PS
B	NB	NS	NS	Z	PS
L	NB	NB	NS	NS	Z

Table 1. Angular velocity rules

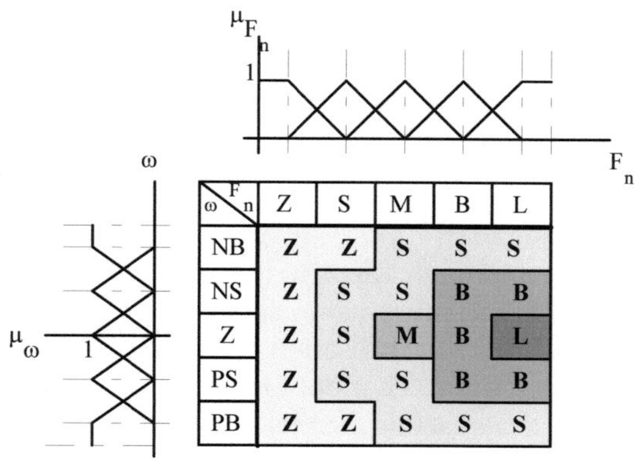

ω \ F_n	Z	S	M	B	L
NB	Z	Z	S	S	S
NS	Z	S	S	B	B
Z	Z	S	M	B	L
PS	Z	S	S	B	B
PB	Z	Z	S	S	S

Table 2. Linear speed rules

An example of such fuzzy rules is:

If (R_n is *Large*) **and** (L_n is *Large*) **then** (ω is *Zero*) **and if** (f is *Large*) **then** (v is *Large*).

Complete results are detailed in [Hoppenot96] and [Benreguieg97].

2.3 Navigation behaviour

The ω and v control actions produced by the above fuzzy controller handle the robot to avoid the obstacles when it is attracted by the immediate nearest subgoal (SG^k). This latter exercises an attractive force which guides the robot to its destination. The actions (ω_a and v_a) generated by this force are modulated by the inverse of the distance $\left(\left\|R, SG^k\right\|\right)$ between the centre of the robot and the kth subgoal.

$$\text{If } \left(\left(L_n \notin [0.2;0.4]\right) \text{ or } \left(R_n \notin [0.2;0.4]\right) \text{ or } \left(F_n < 0.2\right)\right)$$

$$\text{then: } \quad \omega_a = 0 \quad \text{and} \quad v_a = 1$$

$$\text{else if } \left(\left\|R, SG^k\right\| > D\right)$$

$$\text{then: } \quad \omega_a = \frac{C_{at}}{\left\|R, SG^k\right\|}\theta_a \quad \text{and} \quad v_a = 1 - \omega_a$$

$$\text{else: } \quad \omega_a = 0{,}5\ .\ \theta_a \quad \text{and} \quad v_a = 1 - \omega_a$$

The setpoints V and Ω applied to the robot result of a linear combination between the obstacles avoidance and the subgoal attraction.

$$\text{If } \left(\left\|R, SG^k\right\| < D\right) \text{ or } \left(\left\|R, SG^{k-1}\right\| < D\right)$$
$$\text{then } V = Min(v, v_a).V_{\min} \ (m/s)$$
$$\text{else } \quad V = Min(v, v_a).V_{\max} \ (m/s)$$
$$\Omega = \beta\ (\omega + \alpha\omega_a) \quad (\text{rd/s})$$

where Cat, α and β are coefficients adjusted by experimentation to get the best trajectory generation.

2.4 Localisation

Planning and navigation (in the sense of following the planned trajectory) are possible only if the robot has information about its localisation. Localisation methods are divided into two families. Relative localisation consists in computing present position taking into account the previous one and the robot displacement. This method is easy to implement in real-time, but its main drawback is that its error is not bounded and tends to grow with time. Absolute localisation is based on exteroceptive perception and knowledge of the environment. It is performed in 4 steps: (i) data acquisition (here with a camera), (ii) primitive extraction (here segments), (iii) 2D-3D Matching and (iv) position computing. The major part of our work was done on the last two points.

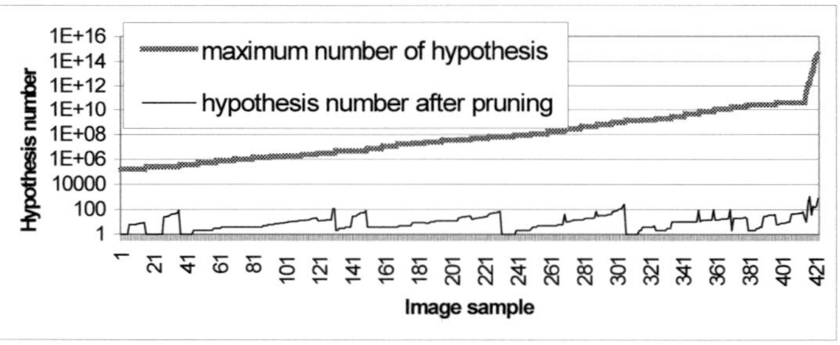

Fig. 1. Pruning performance with unary and binary constraints

Concerning matching methods, they can be classified in two groups: methods which search a solution in the "correspondence space" such as alignment ([Ayache86], [Lowe87]), geometric hashing ([Lamdan88]) or interpretation tree search ([Grimson90b], [Grimson87]) and those which search in the "transformation space" such as generalised Hough transform ([Grimson90a]). One of the most popular approaches is the interpretation tree search introduced by Grimson ([Grimson90b], [Grimson87]). We have proposed a two-stage method for mobile robot localisation based on a tree search approach and using straight line correspondences. The first stage serves to select a small set of matching hypothesis. Indeed, exploiting some particularities of the context, the sets of image lines and model segments are both divided in two subsets. Two smaller interpretation trees are then obtained. Two different geometric constraints (a unary constraint and a binary constraint) which can be applied directly on 2D-3D correspondences are derived and used to prune the interpretation trees. In the second stage, poses corresponding to retained matching hypothesis are calculated. An error function is used to select the optimal match if it does exist. Figure 1 shows the number of hypotheses to be tested regarding their total number. As the ordinate axis scale in a logarithm one, the method gives very interesting results. Considering that point (iv) is the time consumer, it is very important to reduce as much as possible the number of hypotheses to test. All results can be found in [AitAider05], [AitAider02b] and [AitAider01].

Position computing (step (iv)) is based on Lowe's algorithm ([Lowe87]). We have proposed two main adaptations of this method. First of all, Lowe's method is based on point correspondences. In this application, the image is first segmented into contours. Contours generally correspond to physical elements in the work space, such as edges constituted by intersections between surfaces of the flat. These edges tend to be straight segments. Lines are easier to extract from contour images and their characterisation by polygonal approximation is reliable even in the presence of noise. Partial occlusion (due to the view angle or the presence of non-modelled objects) does not affect line representation parameters. Furthermore, the extremities of the edges that could possibly be considered as point features are not always seen in the image due to the dimension of the flat edges in comparison with their distance to the camera. These reasons make it more prudent to use straight line correspondences. Thus, the 3D model can simply comprise a set of straight segments whose extremities have known co-ordinates in the world frame. The second adaptation concerns the degrees of freedom of the system. Lowe's algorithm works in 6D (three positions and three orientations). In our context, we have only two positions and one orientation as the robot evolves in a 2D environment. That gives the possibility to reduce the number of parameters. The obtained system of equations is still non-linear and contains multiple unknowns. Convergence properties are highly dependent upon the quality of the initial estimate of the solution vector. Many situations unfortunately arise in which the robot is "completely lost" in its environment and has no perception of its actual location. An approach to reducing the effects of non-linearity is to find a way to uncouple some of the variables. The rotation and translation parameters have been uncoupled. An initial estimate of the solution can be found by analytically solving one of these equations. A numerical optimisation by means of least squares using Newton's method is then to be applied. Development can be found in [AitAider02a].

3. Human machine cooperation

3.1 Appropriation and human-like behaviour

According to Piaget, the intelligence is before all adaptation ([Piaget36], [Piaget52]). The functional organization of the living being emerges from the balanced relation which is established between the individual and the environment. This balance is made possible by transformations induced by the characteristics of the environment with which a person interacts. For Piaget, who analyzes the birth of the intelligence in its sensorimotor dimension, the adaptation can break up into two processes.

The first one is the process of assimilation. According to this author, this process is defined as the tendency to preserve a behavior. That is made possible thanks to a certain repetition of the behavior in question which thus is schematized. A scheme constitutes a structured set of the generalizable characteristics of the action which will allow the reproduction of the same action even if the scheme is applied to a new but close situation. These schemes constitute an active organization of the lived experiment which integrates the past. They thus include a structure which has a history and progressively changes with the variety of the situations met. The history of a scheme is that of its generalization but also of its adjustment to the situation to which it is applied. Generalization is conceptualized by the process of assimilation. Concretely, by their proximity of appearance or situation, the use of new objects can be assimilated to pre-existent schemes. The property of differentiation, as for it, refers to the second process responsible for the adaptation called by Piaget accommodation. When the complexity of the situation does not allow a direct assimilation, a mechanism of accommodation builds a new scheme by important modifications of pre-existent schemes. If one takes the example of the acquisition of the manipulation of a stick by the young child ([Piaget36], [Piaget52]), one completely understands the nature complementary of this process with that of the assimilation. In this experiment, a child is placed vis-a-vis a sofa on which a bottle is posed out of range. However within his hand range, there is a stick. Initially he tries to seize the bottle directly, then seizes the stick and strikes with the stick and by chance makes object fall. When the bottle is on the ground, he continues to strike by observing the movements obtained, then he ends up pushing the object with the stick to bring it back towards him. Later, in the absence of the stick, he seizes a book to bring closer the bottle. The child thus, first of all, implemented a scheme already made up - to strike with a stick - but this assimilation of the situation to the scheme does not make it possible to succeed each time. Consequently the scheme gradually will be adapted in order to manage the displacement of the object, until leading to a new scheme: to push with a stick. Lastly, this one will be generalized to other objects, here a book. The same mechanism is applied to the man-machine relation. The development of sensorimotor schemes in the young child is relatively transposable with the situation of the operator having to build schemes of action for controlling the robot. When the machine presents operating modes close to those known by the operator, he builds robot control schemes by an assimilation process based on preexistent schemes. On the contrary, if the machine operating is completely different, the person is obliged to accommodate. It is this principle of adaptation, at Piaget sense, applied to the man-machine relation which we describe as mechanism of appropriation.

3.2 Ergonomic design

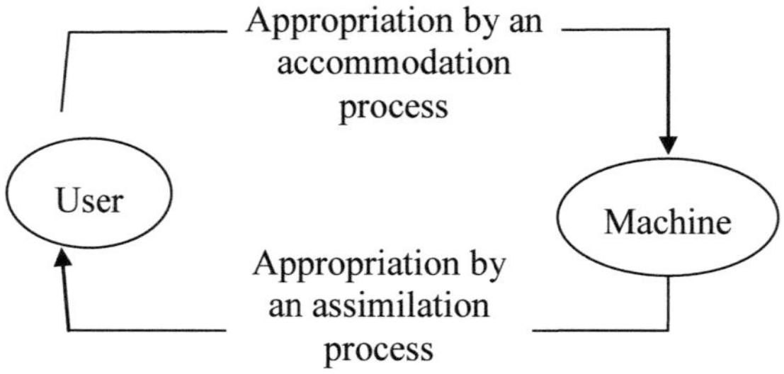

Fig. 2. Application of the Piaget model of adaptation to the Man-machine Co-operation

The preceding considerations expose that the nature of the adaptation of the human to the machine is strongly dependent on the operating difference between these two entities (Figure 2.). If the difference is weak, the adaptation is carried out by a process with a dominant of assimilation, i.e. by the generalization of the initial schemes relevant for the control of the machine. Conversely, if there is a high difference of operating, the operator is confronted with a situation which is completely unfamiliar for him. It is thus the process of accommodation which becomes, for a time, dominating. It leads to the transformation and reorganization of available schemes, which gradually produce new compositions of schemes allowing the reproducible management of the new class of situations. It comes out from these observations that the question of the difference existing between the schemes and initial representations of the operators and, the schemes and representations necessary to control the machine are crucial in the ergonomic design of the human-machine co-operation. Within this framework, two complementary approaches are conceivable.

The first option is to seek to reduce the difference between the existent schemes of the operator and those appropriated to the control of the machine. The approach consists in considering the machine as the prolongation of the motor functions of the user. Thus, when such an assumption is relevant, the user will tend to give his own characteristics and properties to the machine ([Gaillard93]). In this case, an anthropomorphic design seems to be well suited in order to build a directly appropriable tool by the assimilation process. However, in many situations, such a human-machine compatibility is not easily reachable. This is why the second option aims at taking note of the mismatch between schemes necessary to control the machine and those of the user, insofar as the difference is regarded as not significantly reducible. Consequently, the ergonomic design will seek to facilitate the conceptualization of the difference by an accommodation process based on learning.

3.3 Application context

Within the framework of maintenance in residence, a robot able to move, to handle usual objects and to perceive its environment can be an essential complement to the other technological means placed at the disposal of a person with reduced autonomy for her/his safety and in order to ensure other services like the tele-survey, the tele-health, and the social relation… If we take the principle that the robot is semi-autonomous, the user remote controls it (Fi3).

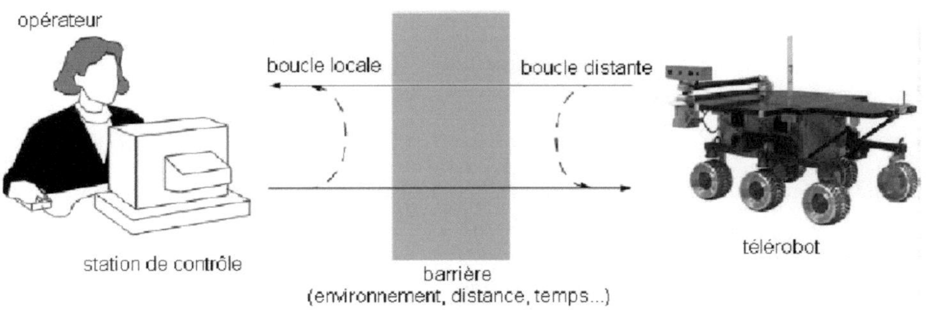

Fig. 3. Remote control situation (adapted from [Fong01])

According to the type of user and use, problems are quite different. The taxonomy defined by the community of the human-robot interaction is a useful tool which makes it possible to determine precisely the nature of the interaction between the robot and its user. Taxonomy gathers the criteria of classification of the interaction in three categories: structural, relational and operational. If one retains only the relevant criteria for our application, this system is composed of user, handicapped or not, and of a robot able to move, to seize and handle objects. In addition, at a given moment, the robot interacts with only one person. Finally the space-time criterion which belongs to the operational category of taxonomy makes it possible to distinguish between three types of tasks. If the robot is remote controlled by the person with reduced autonomy, the robot and the person share the same space i.e. are in the residence of the person. In this case, two situations are possible, the robot is remote controlled either in sight or out of sight. On the contrary if the user is a distant person who remote controls the robot via the Internet, the robot and the user are at different places.

3.4 Control modes

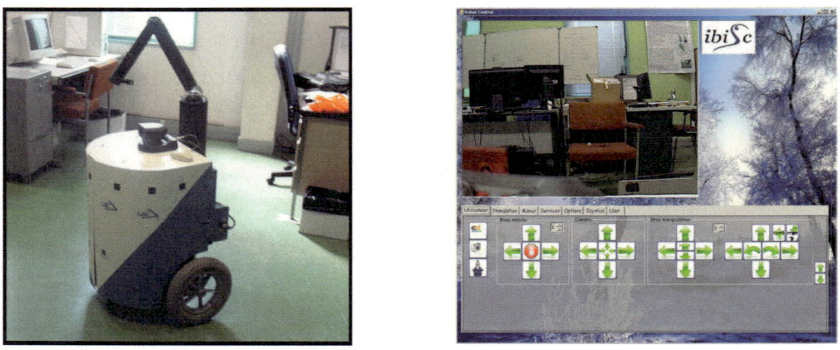

Fig. 4. ARPH system (a) and Human-Machine Interface (b)

The robot of the ARPH project (Figure 4a) is composed of a mobile platform with two driving powered wheels and a MANUS manipulator arm. An ultrasonic sensor ring makes possible to avoid obstacles. The robot is equipped with a pan-tilt video camera. The system - robot and video camera - is remotely controllable via a client/server architecture and a wireless Wi-Fi network.

The user controls the machine from a computer located at a distant place (Figure 4b) by using different control modes.

In the Manual mode the operator controls directly all the degrees of freedom of the robot.

In the Assisted mode, the control of the degrees of freedom of the robot is shared between the user and the machine. This type of mode is the most concerned by a design approach based on appropriation which aims at determining the most efficient way of sharing the control of the robot between the user and the system. As such a type of mode depends on the task and the robot used, it is possible to imagine a large variety of assisted modes.

The key idea is to facilitate an appropriation by the process with a dominant of assimilation (Figure 1) by giving the robot human-like behaviors. The assumption is that the user builds by analogy, more intuitively, a realistic mental representation of the robot. The implementation of behaviors of the human type on the robot rises from the multidisciplinary step made up of the four following stages:

1- Identification of relevant human behaviors, generally perception-action loops

2- Modeling of the behavior's candidates to extract from them the principal characteristics

3- Translation of the model resulting from the neurosciences to an implementable behavior on the robot

4- Evaluation of the mode

We present three examples of assisted-mode design.

3.4.1 Human behaviour candidates

Visio-motor anticipation seems to be a good behavioural solution to palliate the difficulties of space perception and representation. During a displacement, the axis of the gaze of a person systematically anticipates the future trajectory. Indeed, in curve trajectories, head orientation, more precisely gaze direction, of the person is deviated on the inside of the trajectory. This would guide the trajectory by a systematic anticipation of the trajectory direction with an interval of 200 milliseconds ([GRA96]).

An analogy can be made between the direction of human glance and that of the pan-tilt video camera which equips the robot, so we sought to implement on the robot this type of behavior. The foreseen consequence was an improvement of the speed of execution of the movement and the fluidity of the trajectories of the robot. Taking into account the functional architecture of the robot, two implementations of the visual anticipation during a displacement are conceivable: (i) by automation of the anticipatory movement of the camera according to the orders of navigation which the operator sends to the robot or conversely (ii) by automation of the navigation of the robot starting from the orders which the operator sends to the video camera.

3.4.2 Platform mode

In the situation "I look at where I go" that we call in the following "platform mode", the operator directly controls the displacement of the robot. The video camera is oriented only by a reflex action following the trajectory followed by the robot. From the analogy carried out between the human glance and the mobile video camera, the latter is automatically oriented in direction of the point of tangent to the internal trajectory of displacement, i.e. at the place where visual information is most relevant to guide the locomotion (Figure 5). The angle a(t) between the axes of the robot and the pan-tilt video camera must be conversely proportional to the curvature radius of the robot trajectory in order to move the camera towards the tangent point.

By using the trigonometrical properties, $a(t)= \arccos [1- (L/2)/r(t)]$, where $L/2$ is the half-width of the robot.

Due to the fact that the ARPH robot is speed controlled by the user, the radius r(t) is obtained by the ratio between the linear and rotation velocities of the robot.

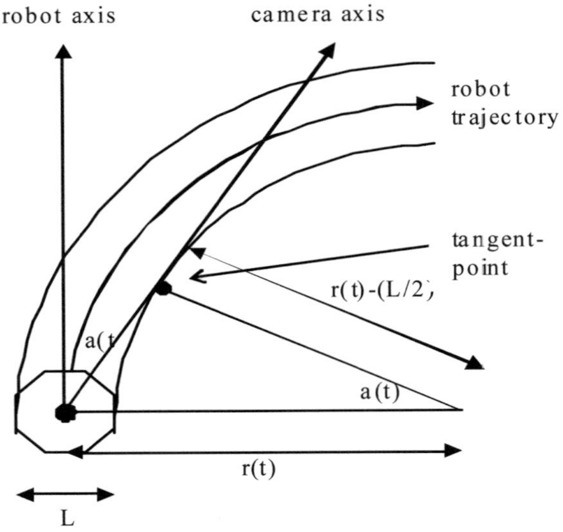

$$a(t) = \arccos(1-(L/2)/r(t))$$

Fig. 5. Implementation of the behavior "I look at where I go"

3.4.3 Camera mode

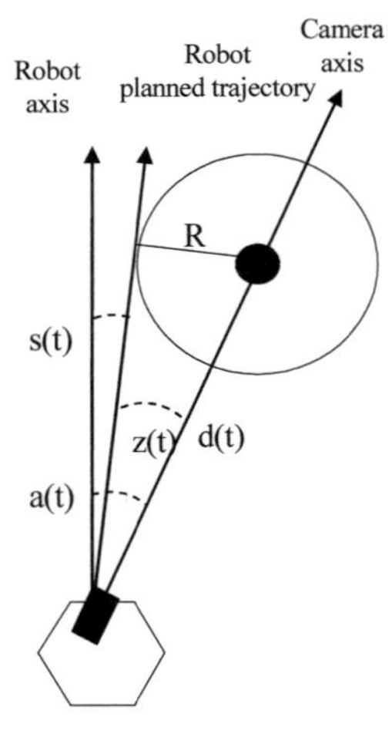

Fig. 6. Implementation of the behavior "I go where I look at"

In the situation "I go where I look at" that we call in the following "camera mode", the operator controls the orientation of the camera directly. The orientation of the robot will be computed from the orientation of the camera. Contrary to the preceding model, the vision is now actively controlled by the user and the displacement of the robot is automated. The model is inspired from the behavior of visual anticipation which consists to fix a reference mark and to maintain it in its visual field in order to describe an ideal trajectory around it. The great advantage of this situation is that it makes it possible the operator to visualize continuously the obstacle nearest to the robot limiting the collision risk.

The angle of navigation of the robot s(t) is defined by the difference between the angle a(t), between the axes of the robot and the pan-tilt video camera, and angle z(t), between the axis of the camera and the tangent to the robot planned trajectory. By using the trigonometrical properties $z(t)= \arcsin[R/d(t)]$ where d(t) is the distance between the robot and the reference mark of navigation and R, the radius of a safety zone around the reference mark. $d(t) = [v(t)/da(t)/dt]\sin[a(t)]$ where v(t) is the linear speed of the robot.

A comparison between manual, platform and camera modes has been realized. The experimental protocol of the evaluation, the results and the data analysis are developed in section 4.

3.5 Robot Learning by the user

The control of the robot is a complex task. It is not enough to give human-like behaviors to the robot so that it is usable. It is necessary to reduce the difference between the mental representation of the instrument that is made by the user and its real use using an accommodation process (cf Figure 1). To assist user for the acquisition of a suitable mental representation of the robot operating, a serie of targeted trainings of progressive difficulty is proposed. The use of a robot simulator presents several well-known advantages to carry out this phase of training. The simulation makes it possible to reproduce, in full safety for the person, situations which can be very varied or not easily realizable in reality. Moreover, the simulator allows a strict control of the experimental conditions and provides exploitable data to judge the degree of acquisition of the skills to be learnt. In addition, it ensures time saving and a reduction of logistic costs of evaluation when the subjects are disabled people. One of the main question is the reproducibility of the results when the user passes from a virtual situation to a real situation, in other words if there is a transfer of skills or knowledge acquired in simulation to the real world. The transfer of training is a relatively important process from an adaptive point of view, "because it is rare that one finds in the life an activity which is exactly that which was learned at the school" ([Lieury04]). However, its implementation is often difficult. "Knowledge is often so related to the context of its acquisition that the individuals meet great difficulties of using them in different contexts" ([Roulin06]). We thus carried out an experiment intended to highlight a positive transfer effect between a situation of training on a simulator of the robot ARPH in its environment, and a situation of transfer consisting in controlling the real robot. One speaks about positive transfer when the effect of former knowledge acquired in a first situation improves the efficiency in a second situation. This efficiency can be translated, for example, by a reduction of the error number, a reduction of the execution time, or an increase of the number of good answers. We formulate the assumption that a preliminary training with the simulator of robot ARPH in a situation whose characteristics are similar to those of the real situation should involve a positive transfer. In short, a dedicated training in simulation should facilitate the use of the real robot.

The experimental protocol of the evaluation, the results and the data analysis are developed in section 4.

3.6 Delay treatment

3.6.1 Related work

In our case, the Internet will be used as the communication medium. That indicates another important problem by adding a transmission delay between the master and the slave computers. The natural reaction to this delay is to adopt a move-and-wait behavior: the operator issues a command and waits for the robot's feedback. This greatly slows the process of controlling the robot. The delay encountered on the Internet has two main sources: the physical distance that the signals have to travel and the network congestion. One can reasonably expect a continuous component (more or less constant, likely corresponding to the physical distance between the remote operator's and the robot's locations) and a variable component (spikes in the delay, only appearing from time to time, likely because of network congestion) [Moon00], [Garcia03].

When the delay's value is not above a certain threshold (usually, around 300 ms [Henderson01] for tasks involving only a video feedback, but it depends on the type of task and on the operator himself), the operator is not going to be disturbed by it. If it exceeds this value, one needs to think of ways to diminish its influence on the operator. One way of helping the operator overcome the influence of the delay is to adapt the video feedback to match the delay. Another way is to reinforce the human-robot cooperation by relying more on the robot's capacities to manage itself. The former category of aids for the remote control (adapting the video feedback), has the advantage of not requiring a very computational-powerful robot. One of the problems with it is that, for certain types of video feedback aids, it requires some a priori knowledge of the environment and for the robot to be the only thing moving (or for the environment to evolve very slowly).

An aid that seems to give very good results is to use a predictive window [Baldwin99]. In this case, the camera acquires a 360° image and only a portion of this image is presented to the operator. The purpose of this is to reduce the network traffic, as only the selected portion will be transmitted to the operator. It is also possible to think of the image as a texture, so that when the camera is moving, the texture that was represented by the initial image will be deformed as to match the new viewpoint ([Cobzas05]). If it isn't too complicated to create a model of the environment and if the environment isn't expected to change (or to do it slowly, so as to have time to update the model), we could think of using a representation in virtual reality [Bares97]. In this way, the robot will be controlled in the virtual world. If creating the whole model isn't possible, using a representation of the robot (or parts of it - such as its end effector) could be used to help alleviate the problems induced by an unreasonable delay. Such is the case of [Friz99], who uses a mechanical arm over a workbench, whose task is to create different structures using wooden blocks. In this thesis work, the operator can use three cameras located around the workbench and a camera that's attached to the arm's base. Visual cues concerning the future position and orientation of the robotic arm are given. They are represented in such a way that their interference with the real image is minimal (according to the "less is more" principle).

3.6.2 Proposed solution

What we aim for in our system is to either give the impression that the communication between the remote operator and the robot happens with virtually no delay (just the minimal time necessary to process the information) or to allow the operator to see the position where the robot should be if there was no delay.

As already mentioned, the communication between the remote operator and the robot will pass through the public Internet. This means that it is possible that we will have to deal with the influence of the delay that will exist between the robot and the operator's computer.

The article that served as the basis of the moving window aid is [Baldwin99]. Because we don't have a panoramic camera (60° field of view), if the robot will turn too much to the side, a black area will be used instead of the image that should have been there. An improvement on this kind of aid is that, when the robot advances, a portion of the image (central, if the robot moves straight ahead) will be zoomed in on the display interface. This will give the impression of the robot moving forward for the operator. If the robot is to move backwards, the selection window will enlarge (while, as in the previous case, the image will be scaled as to fit into the display interface). A scaling factor can be applied just before displaying the modified image, as to be able to control the sensation of turning/advancing. All these modifications will be realized on the operator's side of the remote control system. For the deploying stage, we are currently investigating how to synchronize the clocks of the robot and of the remote operator's computer using the network time protocol [Mills90]. According to [Elson02], the maximum difference between the two clocks will not be superior to 2 ms, which is good enough for our case, when the operator is disturbed only when the delay exceeds 300 ms [Henderson01].

A second type of aid is to use a simplified model of the robot, which will be presented on screen to the operator when, because of the delay, the image displayed is not the one that he should be seeing. This method has its roots in [Friz99]. For our case, we don't always display the virtual robot (called a phantom from now on). The phantom will appear only when the distance between the position that the operator currently has for the robot and where the robot should be (in the absence of delays) goes above a certain value (0.2 m for the tests).

4. Evaluations

4.1 Comparison of human-like control modes

The evaluations concern the assisted modes presented in §III.4. The hypotheses of our work were the following ones. Firstly, a situation in which the pan-tilt video camera is mobile and points towards the future trajectory of the robot, should lead to better performances in terms of control of the trajectory than a situation in which the camera is fixed and always pointing along the axis of the machine. Secondly, by reference to works evoked above, a strategy of the type "I go where I look at" will lead to a more optimized trajectory than "I look at where I go".

4.1.1 Experiment results

Three command modes have been tested, two with anticipation -camera mode and platform mode- one without anticipation. This last mode corresponds to a fixed camera, aligned with the robot axis, and a manual control of the displacement of the platform.

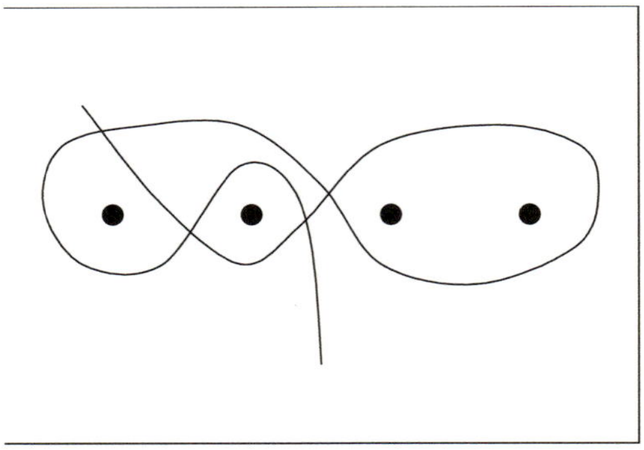

Fig. 7. Schematic representation of the trajectory.

The hypothesis is that in "camera" control will be better than in "platform" mode. The "fixed" mode is used as a reference. The operator has to realise with the robot the trajectory given in Figure 7. The criteria of comparison are performance parameters: trajectory execution time, collision number, stop number and a behavioural parameter: trajectory smoothness.

About collisions (Figure 8), anticipation conditions ("camera" mode plus "platform" mode) present significantly less collision than "fixed" mode ($p<0.01$). There is no significant difference between "camera" mode and "platform" mode.

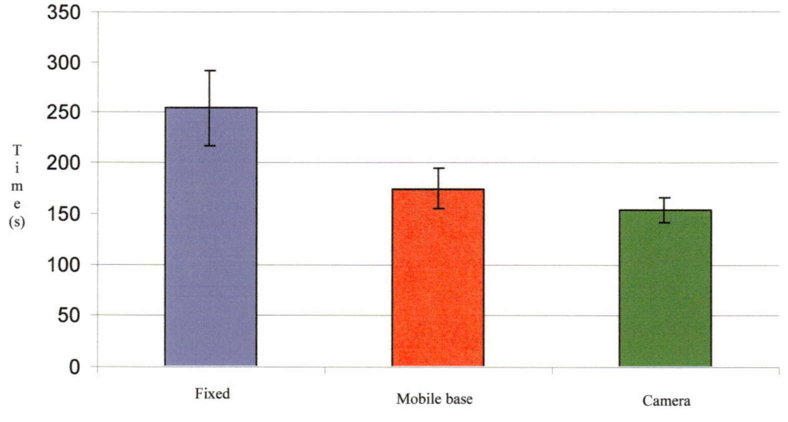

Fig. 8. Mean time of execution.

Concerning number of collisions (Figure 9.), anticipation condition have significantly less collisions than "fixed" condition ($p<0.03$). But there is no significant difference between "platform" mode and "fixed" mode. There is also no significant difference between "camera" mode and "platform" mode.

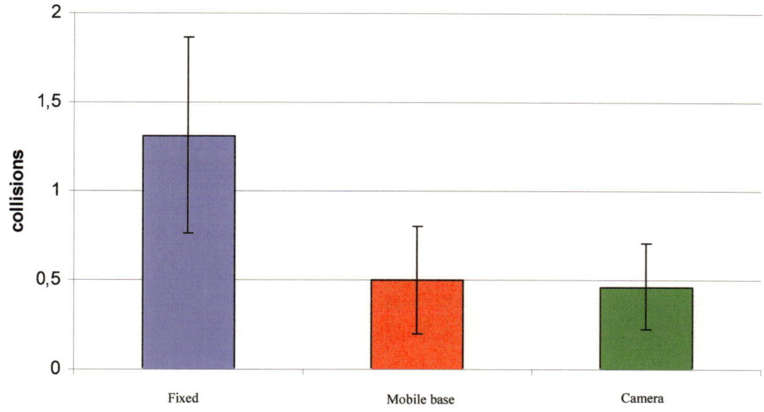

Fig. 9. Mean number of collisions.

About number of stops, both anticipation conditions are very significantly better than "fixed" condition.

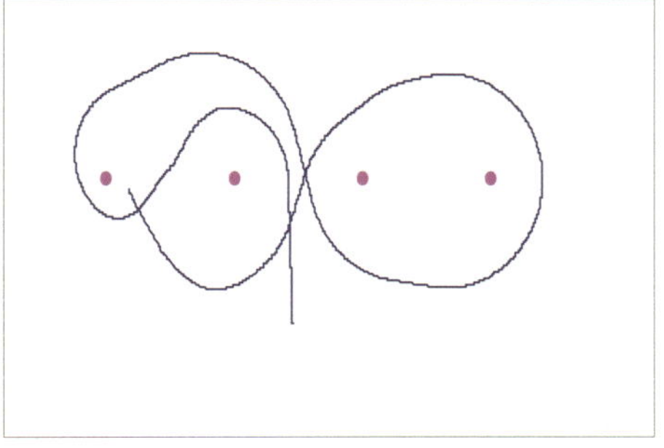

Fig. 10. Trajectory in anticipation mode.

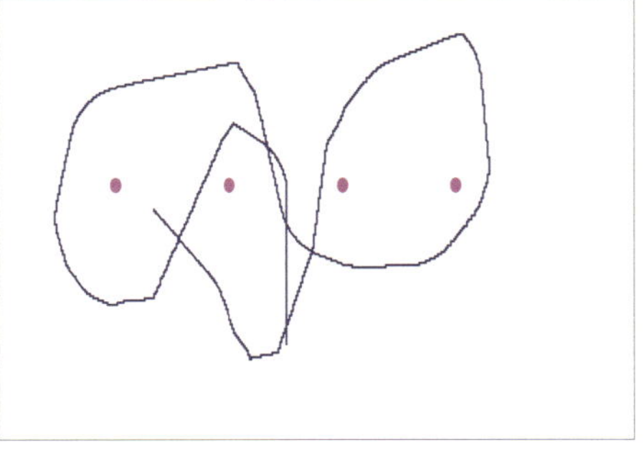

Fig. 11. Trajectory in "fixed" mode.

Let us now examine behavioural parameters. Figure 10 shows a trajectory in anticipation condition. Figure 11 shows a trajectory in "fixed" condition. The second one is more angular than the first one. [Peruch99] proposes a solution to quantify this. Number of occurrence of radius of curvature (r=v/w, v: linear speed, w: rotation speed) is represented in Figure 12 X axis is expressed in logarithm of radius of curvature, Y axis is the occurrence percentages of these radius of curvature.

A small radius of curvature (log(r)<0) represents a small linear speed and big rotation speed, which corresponds to angular trajectory. An important radius of curvature (log(r)>0) represents a smooth trajectory. Figure 12 shows an important occurrence of radius of curvature around log(r)=0 in all conditions. That corresponds to mean radius of curvature. But, around log(r)=-2, which corresponds to about only rotation, this number is significantly higher in "fixed" condition than in "platform" condition, which higher than in "camera" condition. So, anticipation conditions are better than "fixed" condition, but also anthropomorphic condition ("camera") is better than non anthropomorphic condition ("platform").

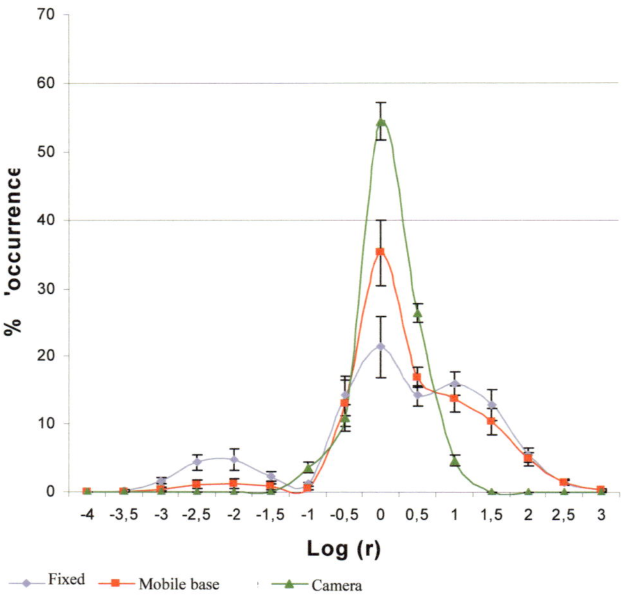

Fig. 12. Occurrence percentages of radius of curvature.

4.1.2 Discussion

Two kinds of parameters have been studied to compare the three control modes. Performance parameters (trajectory execution time, number of collisions, stop number) show that both visual anticipation modes ("camera" and "platform") give better results than "fixed" mode.

The trajectory smoothness which is a behavioural criterion confirms the previous results. The occurrence of radius of curvature shows that there is significantly less small radius of curvature, corresponding to pure rotation, in "camera" mode than in "platform" mode". The control is smoother in "camera" mode than in "platform" mode.

4.2 Learning transfer

As shown in section 3, the evaluation intended to highlight a positive transfer effect between a situation of training on a simulator of the robot ARPH in its environment, and a situation of transfer consisting in controlling the real robot. To be interested in the transfer during the development of a preliminary training makes it possible to know the methods likely to allow a positive transfer, those which will facilitate the acquisition of the abilities and knowledge, but also to know the methods capable of involving a negative transfer and thus to obstruct the operator at the time of his passage in situation of transfer, and which can induce even dangerous unsuited behaviors in real situation.

4.2.1 Experimental Protocol

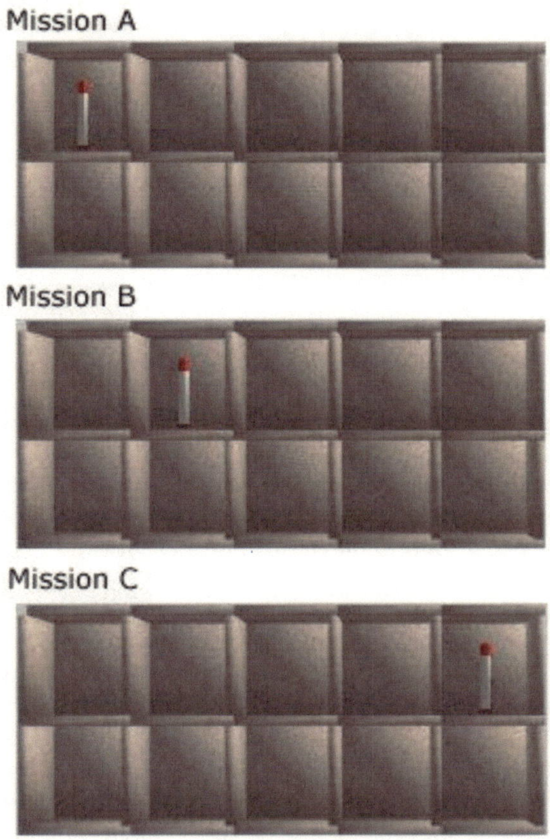

Fig. 13. Pen position in the pigeonhole for each mission.

The subjects had to carry out missions consisting of moving the robot and seizing a pen posed in racks, the robot being placed 3m in front of the racks. The subjects had to carry out three different missions corresponding to three positions of the pen in the racks. The experiment was carried out by three groups of 10 subjects. Since this robot is likely to be used by all, it did not seem necessary to us to carry out sampling, in particular concerning the sex, the age, or competences of each one. The population was heterogeneous and the subjects were not informed of the objective of the experiment. The 30 subjects were divided into three groups: an experimental group, and two control groups. There are then three conditions. The Virtual-Virtual (V-V) and Reality-Reality (R-R) conditions correspond to the two control groups which carry out 12 missions. The Virtual-Reality (V-R) condition corresponds to the experimental group, which carries out 6 missions in virtual situation, then 6 missions in real situation. The subjects were informed that after the six missions on the simulator, they would carry out six new missions with the real robot. As in the simulation situation the gripper cannot be opened or closed, the task was then to position the gripper near the pen (Figure 13). The same constraint has been given to subjects in the real situation. There were three different missions corresponding to three positions of the pen in the pigeonhole. For the mission A the pen was placed in the top left compartment, for the mission B the pen was placed in the second top left compartment, for the mission C in the top right one (Figure 14). These different positions of the pen have been selected in order to require robot side movements for the missions A and B. The order of the missions during the experiment has been decided according to a Latin square to avoid an effect of order and series. Each of the subjects had to realize four times each mission according to six possible different series. The experiments were preceded by an explanation on the robot use as well as a demonstration either with the real robot or with the simulator, depending on the condition. The experience was then in two phases: a six missions learning phase followed by a six missions stand-alone use. During the learning phase, the investigator answered any questions. From the seventh mission, when began the second phase, subjects remote controlled the robot without any help of the investigator.

4.2.2 Data and method of data analysis

The data collected in order to evaluate the performances of the subjects during tests are the execution time of the mission, the number of stops, the number of reversing, and the final orientation of the robot. Four variables must be characterized. The time execution is a continuous data bounded on the left by zero, which supposes it respects a Gamma distribution. The number of stops and the number of reversing are discrete data bounded on the left by zero, which supposes they respect a Poisson distribution. Finally the orientation is a continuous and not bounded data respecting a Normal distribution. ANOVA is usually used for data analysis. However such an analysis supposes as a preliminary that the distribution is normal. In our case, only one of the four variables respects this condition. So for analyzing the data we used another method ([Raftery95]). The principle is to choose among several models that which describes the data as well as possible. We then compared the BIC (Bayesian Information Criterion) which enables to know the probability that the model is true. The model for which the BIC is the smallest BIC is the most probably true. Indeed, the BIC takes into account the likelihood of the model compared to the data i.e. its adjustment with the data, as well as the number of unknown parameters. The more the adjustment is close and the number of unknown parameters low, the smaller the BIC is. For each variable, we compared five models. The analysis of the BIC was carried out with the software of statistics "R". Once the most probably true model is found using the BIC, a graph is obtained, the curve of regression, representing the data according to this model. We can then directly see if there is a learning effect. If the curve is a straight line then there is no effect, if the curve is inclined then there is an effect.

4.2.3 Results

4.2.3.1 Execution time of the mission

As we have seen, data obtained for this variable are continuous and bounded on the left by zero, which supposes it respects a Gamma distribution. In the R software, for a gamma the default selected regression function law is a reverse function. We have also tested models using the exponential function. Here are the models tested for this variable:

- The M0 model assumes that there is no learning effect; the M1 model assumes that there is a learning effect that is the same for all subjects;
- The M2 model assumes that there is a different effect of learning between those who have learned in virtual situation - the V-V and V-R groups - and those who have learned in real situation - the R-R Group -;
- The M3 model assumes that there is a different effect of learning for each group;
- The M4 model assumes that there is a different learning for each group, with a break between the first six trials and the last six trials.

Table 1 shows the BIC for these different models tested with the gamma table and using the two types of function:

Model	Reverse function	Exponential function
M0	9211.728	9211.728
M1	9076.422	9094.752
M2	9068.998	9088.089
M3	9050.025	9075.586
M4	**9037.607**	9060.251

Table 1. BIC for *the execution time variable.*

Thus, it can be seen that the smallest BIC is the M4 model according to a reverse function. The model that is the most probably true is that which assumes a different learning between groups with a break effect between the two phases. We see on the graph (see Figure 14) that learning is not made in the same way for the three groups. Between the two phases there is therefore a break. The V-R group was the fastest group during the first phase and became the slowest one in the second phase. For the other two groups there is also a slight increase in the time at the beginning of the second phase. Then times decrease in the second phase for all groups, however less quickly than during the first phase.

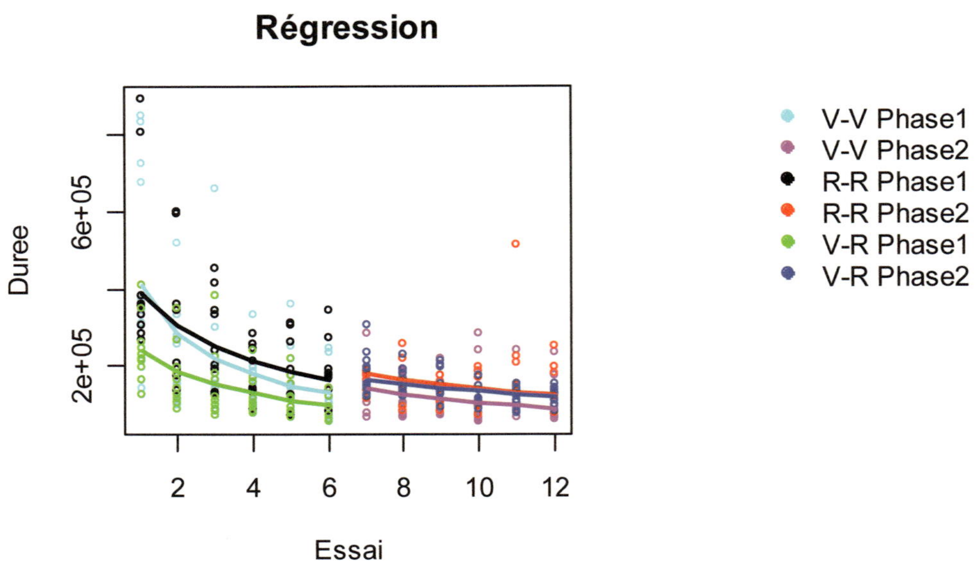

Fig. 14. Regression curve for the execution time. Time is in ms.

4.2.3.2 Robot orientation at the end of the mission

The data of this variable are continuous and not bounded respecting a Normal distribution, so it is possible to perform an analysis of variance. However we have privileged the BIC analysis as a first step to retain, as previously, the model probably true. The analysis of variance has been used later to confirm or not the choice of the chosen model. The models which have been tested for this variable are not quite the same as for the execution time variable. There are also five models:

- The O0 model assumes that there is no learning effect;
- The O1 model assumes that there is a learning effect that is the same for all subjects;
- The O2 model assumes that there is a different effect of learning for each group;
- The O3 model assumes that there is a learning which is the same for all subjects but different between the two phases;
- The model O4 effect assumes that there is a different learning for each group with a break between the first six trials and the last six trials.

The data being continuous and not bounded a Normal distribution has been selected for the BIC analysis. The smallest BIC is the O0 model which assumes the absence of a learning effect. A variance analysis also shows us that the best model is the O0 model. The regression curve of this model is a straight line (see Figure 15), denoting that there is no effect.

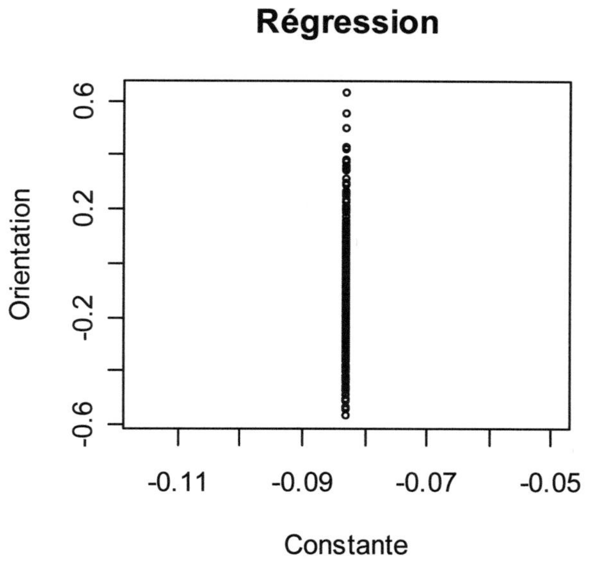

Fig. 15. Regression curve for the Orientation.

4.2.3.3 Number of stops during the mission

Data obtained for this variable are discrete data bounded on the left by zero but not on the right which supposes they respect a Poisson distribution. So it has been chosen to use a Poisson law to perform the analysis of the BIC. Models tested here are the same as for the execution time:

- The S0 model assumes that there is no learning effect;
- The S1 model assumes that there is a learning effect that is the same for all subjects;
- The S2 model assumes that there is a different effect of learning between those who have learned in virtual situation - the V-V and V-R groups - and those who have learned in real situation - the R-R Group -;
- The S3 model assumes that there is a different effect of learning for each group;
- The S4 model assumes that there is a different learning for each group with a break between the first six trials and the last six trials.

The smallest BIC is that of model S3 which is therefore the model probably true. The learning effect differs depending on the group (see Figure 16).

The observation of the curve of regression (Figure 16) shows that the V-V group carries out a greater number of stops during the first tests. This number progressively decreases for finally being lower than those of the two other groups. The R-R and R-V groups are relatively equivalent, the number of stops of the V-R group being slightly higher at the beginning and a little lower during the last trial.

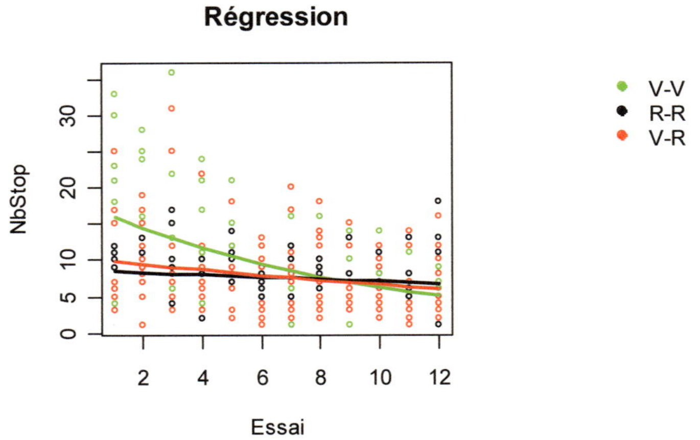

Fig. 16. Regression curve for the number of stops .

4.2.3.4 Number of reversing

Data obtained for this variable are discrete data bounded on the left by zero but not on the right which supposes they respect a Poisson distribution. So it has been chosen to use a Poisson law to perform the analysis of the BIC. Models tested here are the same as for the execution time:

- The A0 model assumes that there is no learning effect;
- The A1 model assumes that there is a learning effect that is the same for all subjects;
- The A2 model assumes that there is a different effect of learning between those who have learned in virtual situation - the V-V and V-R groups - and those who have learned in real situation - the R-R Group -;
- The A3 model assumes that there is a different effect of learning for each group;
- The A4 model assumes that there is a different learning for each group with a break between the first six trials and the last six trials.

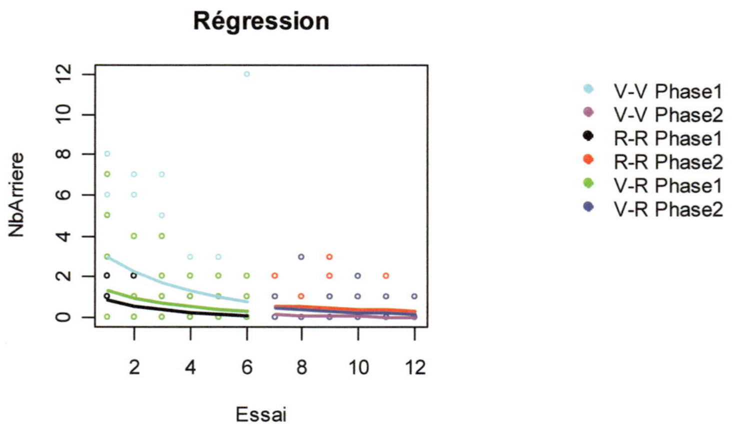

Fig. 17. Regression curve for the number of reversing

The A4 model presents the smallest BIC which assumes that learning is different according to the groups and that there is a break between the first and the second phase. The V-V Group presents a number of reversing slightly more important during the first phase than the R-R and V-R groups. During the second phase, the V-V group shows the lower number of reversing (see Figure 17).

4.2.4 Discussion

In summary, with regard to execution time but also reverse gear we noted a positive transfer effect of learning between the situation of simulation and the use of the physical robot. Simulation makes it possible to the operators for acquiring the necessary abilities as well as transferring them for the use of the physical robot. Users learned how to control the robot, how to use the control interface and how to find there the indicators relevant with a correct use of the robot. These results confirm our assumption and justify a preliminary training in simulation before really using the robot. Indeed, only six tests in virtual situation made it possible for the subjects to be as effective during their first use of the physical robot as the subjects being already at their seventh test in this situation, and this as well with regard to time, the number of stops, the number of reverse gear and the orientation of the robot.

More precisely, concerning the execution time, one observes indeed a transfer positive effect. The results of V-R group show that the operators, when they perform the task in real situation after performing six trials in virtual situation are as fast as the operators of R-R group. Despite the fact that at this second phase the operators of the V-R group are beginners for the real situation the results indicate that they are as good as the R-R group which can be considered as experts since they have carried out six trials moreover in this situation. One can also note that they are faster than the beginners in real situations. One observes that training on the robot simulator is an important time-saver and makes it possible to reach the same level of skill in real situation as an "expert" of the R-R group. The training on simulator thus seems to make it possible for the operators to acquire abilities for the use of the robot which they could transfer when they passed in real situation. They have then the capacities to use the robot as from their first trial in real situation as well as operators having already certain experience. The training on simulator enables them to be effective from the beginning with the real robot. One can also note that, as operators of V-R group take the time in real situation than operators in R-R group, the operators of V-R group do not seem to take more risks than the operators of R-R group.

Regarding the variable "Orientation", the results indicate that there is no effect of training. It thus seems that in a "natural" way the subjects more or less drive the robot in front of the compartment of the pigeonhole which contains the pen. One cannot then observe a transfer of training between the virtual situation and the real situation. It still should be noted that there is not negative transfer. It is the same for the variable "Number of stops".

Lastly, concerning the variable "number of reversing", one can observe a positive transfer. When the subjects of the V-R group perform the task in real situation, one notices a light increase in the number of reversing compared to the first phase. But they then have performances equivalent to those of R-R group and even slightly better.

4.3 Delay treatment

For the tests, we used a direct connection to the robot with a simulated delay. The total delay between the robot and the operator was set at 1 s. Each aid was evaluated individually (e.g., obstacle avoidance was disabled), as to point out only its influence on the remote control system.

The protocol was conceived as to enable comparing the efficiency of two aids to the remote control (moving window and phantom robot) when a delay is present. Each of these aids represents a condition that we're going to compare against a control condition (controlling the robot without having any aid enabled).

Each participant had to pilot the robot for a total of six times. Each participant was presented with two of the three conditions (no aid, moving window, phantom robot), the third condition (phantom robot) becoming available after the beginning of the evaluations.

The fact that each test subject had six opportunities to control the robot could pose some learning issues, but we undertook the following steps to limit them: each time the subject controls the robot, it's on a different path. Six basic paths had been defined. Each of these paths consists of a series of three turns, left or right, linked in such a way that each path is different from the others. For each participant, the paths were presented in a different order. Markers were glued on the ground, to ensure that the same path will suffer no deviation from one participant to the next. The order in which the different conditions were presented to the subjects was counter-balanced, to make sure that the learning effects are equally distributed between the conditions.

During the experiments, the robot was introduced to the participant and his first task was to control the robot (using direct vision, if so he wished) with no delay present. Next, the subjects were taken to another room, with no direct visual contact with the robot. In order to pilot the robot, they had to rely on its camera. Before controlling the robot on each path, the starting and ending points, as well as a direct view of the path were presented to the subjects, in order to avoid the subjects feeling lost and spending the first tens of seconds trying to locate themselves in the environment. The different aids or conditions were not explained to the subject, as to not bias his manner of interacting with the command interface. Three dependent variables were measured during each stage: total time for a path, and two kinds of collisions: the number of errors (collisions) due to a bad approximation of distances and the number of turning errors. The collisions due to distance errors are tied to an imperfect evaluation of the distance that is in front of the robot and are registered when the robot hits an object while moving straight forward. The collisions due to turning errors happen when a turn is begun at a bad moment and a collision ensues. We suppose that an error happens as soon as the robot touches an object from the environment.

The first results are based on statistics applied to the time (T) necessary to the subject to go through the labyrinths (in seconds), and to the numbers of collisions done while turning (C). They were not applied to the number of collisions done while going forward, as there too few of them for any results to be significant. The three conditions (control, zoom-like, phantom-like) will be compared using a one-way analysis of variance (ANOVA). There are 42 observations for condition 1 (no aid), 38 for condition 2 (moving window), and 12 for condition 3 (phantom robot). In order to test the homogeneity of variances, a Levene Statistic is used, which results in a significance of 0.198 for t, and of 0.012 for m. The homoscedasticity seems less important for t, but the ANOVA is quite robust with different variances as long as the samples have roughly the same size. This means that the results of the ANOVA are less reliable for the third condition. Concerning the duration of the paths (T), in condition 1, $\bar{x}$=85.00 s, sd=32.107, and the results are comprised between 45 s and 165 s. In condition 2, $\bar{x}$=65.05 s, sd=26.262, and the results are comprised between 37 s and 165 s. In condition 3, $\bar{x}$=70.75 s, sd=21.760, and the results are comprised between 31 s and 100 s. The F of Fischer is F=4.985, p= 0.009, meaning the groups don't seem to be equivalent. Student test gives t=-3.053, p=0.003 between condition 1 and 2, t=-1.781, p=0.086 between condition 2 and 3, and t=-0.751, p=0.461 between condition 1 and 3. By using a one-way ANOVA, we find that the mean difference is significant only between condition 1 and condition 2 (Mean Difference = 19.947, p=0.009). These results show a strong effect of time reduction when using the zoom, 23.4% on the average. The other results seems to show nearly no difference at all between condition 1 (no aid) and condition 3 (phantom robot), and a little effect between condition 2 and condition 3 (though not being significant), as if condition 1 and condition 3 were quite equivalent.

With regard to the number of collisions C made, in condition 1, $\bar{x}$=0.33, sd=0.477, and the results are comprised between 0 and 1. In condition 2, $\bar{x}$=0.74, sd=0.860, and the results are comprised between 0 and 3. In condition 3, $\bar{x}$=0.67, sd=0.888, and the results are comprised between 0 and 3. The F of Fischer is F=3.391, p= 0.038, meaning the groups don't seem to be equivalent. Student test gives t=-2.558, p=0.013 between condition 1 and 2, t=1.250, p=0.233 between condition 2 and 3, and t=-0.241, p=0.813 between condition 1 and 3. By using a one-way ANOVA, we find that the mean difference is significant only between condition 1 and condition 2 (Mean Difference = 0.404, p=0.039). These results mean that the zoom-like condition seems to provoke 0.404 more collisions, i.e. 122% of collisions done without zoom. This number may seem important, but it must be remembered that the scale of the collisions is quite small.

5. Conclusion and future works

This paper proposes an original approach for a better remote control of an assistive robot. The main idea is that an autonomous robot is not suitable for that kind of tasks. Two intelligent entities, the human operator and the robot, cooperate to achieve the desired missions. To make this cooperation fruitful, the first step is to give to the robot capacities of perception, decision and action. Section 0 deals with this point, giving the robot capacities of trajectory planning, obstacle avoidance and localisation. Among these capacities, obstacle avoidance appears to be the most important one in remote control, as the other two are generally well performed by the user. Nevertheless, trajectory planning and localisation are still important robotics issues.

The way we deal with human machine cooperation is the main originality of our approach. Indeed, the reference to Piaget appropriation theory is a very interesting angle of view for human machine cooperation, which is totally original in this scientific community. The idea is to make the robot as friendly as possible to favour assimilation process. The part of its behaviour that is not natural for the user can be learned by the user through accommodation process, which is more difficult but sometimes the only way of appropriation. Keeping that in mind, we proposed different control modes. Evaluation results show that natural behaviours, meaning behaviours easily understandable by the user, lead to better performances than the others. The same idea has been followed concerning delay treatment. In that case, feedback information to the remote operator is presented as if the movement of the robot would be realised without delay. The robot must have autonomy capacities to make the real movement safe.

We also have developed a simulator of our robot. That offers two advantages particularly interesting in the context of the assistance to the person in loss of autonomy: time saving and training in full safety for the person. In addition, it allows a drastic reduction of logistical costs of training and solves the problem of the low availability of the disabled. This allows to save time with regard to the training of the operators. Indeed, the beginners loose less time to achieve the mission in virtual situation than those in real situation. However, the same number of tests gives an equivalent level to the operators whatever the situation. A formation with simulation thus seems to be as effective as a formation with the real robot, while taking less time. The use of the robot by beginners involves risks. The results of our experiment show that the use of simulation makes it possible to reach a level of expertise equivalent to that of people trained with the physical robot, while avoiding these risks. At the time of the training, in simulation as in real situation, errors can be made, for example the robot or the manipulator can run up against obstacles. However, the consequences are not the same ones for both situations. These errors do not have any consequence, from a material point of view, in simulation, contrary to the real situation for which the same errors can damage the robot. Moreover one knows that the errors can help with the training, allowing to learn what one should not do. Simulation thus makes it possible to the users to make virtual errors, teaching them what it is necessary to avoid making and not to make these errors in real situation again. In addition, making errors in simulation should harm less the confidence of the operators in their capacities to control the robot, contrary to the real situation in which an error has a “cost”. For quadriplegic people who will have perhaps little confidence in their capacity to control such a system, simulation can enable them to acquire this self-confidence, and not to lose it if they make errors.

6. References

[AitAider01] O. Ait Aider, P. Hoppenot, E. Colle : "Localisation by camera of a rehabilitation robot" - ICORR 7th Int. Conf. On Rehab Robotics, Evry, France, pp. 168-176, 25-27 avril 2001.

[AitAider02a] Omar Ait-Aider, Philippe Hoppenot, Etienne Colle : "Adaptation of Lowe's camera pose recovery algorithm to mobile robot self-localisation" - Robotica 2002, Vol. 20, pp. 385-393, 2002.

[AitAider02b] O. Ait Aider, P. Hoppenot, E. Colle: "A Model to Image Straight Line Matching Method for Vision-Based Indoor Mobile Robot Self-Location" - In Proc. of the 2002 IEEE/RSJ international Conference on Itelligent Robots and Systems, IROS'2002, Lausanne, pp. 460-465, 30 September - 4 October 2002.

[AitAider05] Omar Ait Aider, Philippe Hoppenot and Etienne Colle: "A model-based method for indoor mobile robot localization using monocular vision and straight-line correspondences" - Robotics and Autonomous Systems, vol. 52, p. 229-246, 2005

[Ayache86] N. Ayache and O. Faugeras and O. D. Hyper – "A New Approach for the Recognition and Positioning of Two-Dimensional Objects". IEEE Transactions on Pattern Analysis and Machine Intelligence, 8(1), 1986, 44-54.

[Baldwin99] J. Baldwin, A. Basu, and H. Zhang. Panoramic video with predictive windows for telepresence applications, 1999. International Conference on Robotics and Automation.

[Bares97] J. Bares, and D. Wettergreen. Lessons from the Development and Deployment of Dante II, 1997, Field and Service Robotics Conference.

[Benreguieg97] M. Benreguieg, P. Hoppenot, H. Maaref, E. Colle, C. Barret: "Fuzzy navigation strategy : Application to two distinct autonomous mobile robots" - Robotica, vol. 15, pp. 609-615, 1997.Obstacle avoidance

[Cobzas05] D. Cobzas, and M. Jagersand. Tracking and Predictive Display for a Remote Operated Robot using Uncalibrated Video, 2005. ICRA 2005.

[Elson02] J. Elson, L. Girod, and D. Estrin. Fine-grained network time synchronization using reference broadcasts. *ACM SIGOPS Operating Systems Review*, 36(1):147-163, 2002.

[Fong01] Fong, T., & Thorpe, C. (2001). Vehicle teleoperation interface. *Autonomous Robots, 11*, 9-18.

[Friz99] H. Friz. Design of an Augmented Reality User Interface for an Internet based Telerobot using Multiple Monoscopic Views, 1999. Diploma Thesis.

[Gaillard93] Gaillard, J.P. (1993). Analyse fonctionnelle de la boucle de commande en télémanipulation. In A. Weill-Fassina, P. Rabardel & D. Dubois (Eds), *Représentations pour l'action.* Toulouse : Octares.

[Garcia03] C. E. Garcia, R. Carelli, J. F. Postigo, and C. Soria. Supervisory control for a telerobotic system: a hybrid control approach. *Control engineering practice*, 11(7):805-817, 2003.

[Grasso96] Grasso, R., Glasauer, S., Takei, Y., & Berthoz, A. (1996). The predictive brain : Anticipatory control of head direction for the steering of locomotion. NeuroReport, 7, 1170-1174.

[Grimson87] W. E. L. Grimson and T. Lozano-Perez – "Localizing Overlapping Parts by Searching the Interpretation Tree". IEEE Transactions on Pattern Analysis and Machine Intelligence, 9(4), 1987, 469-481.

[Grimson90a] W. E. L. Grimson and D. P. Huttenlocher – "On the Sensitivity of the Hough Transform for Object Recognition". IEEE Transactions on Pattern Analysis and Machine Intelligence, 12(3), 1990, 255-274.

[Grimson90b] W. E. L. Grimson – "Object Recognition: The Role of geometric Constraints". MIT Press, 1990.

[Henderson01] T. Henderson. Latency and user behaviour on a multiplayer game server, 2001. 3rd International Workshop on Networked Group Communications (NGC).

[Hoppenot96] P. Hoppenot , M. Benreguieg, H. Maaref., E. Colle. and C. Barret: "Control of a medical aid mobile robot based on a fuzzy navigation" - IEEE Symposium on Robotics and Cybernetics, Lille, France, pp. 388-393, july 1996.

[Kanal88] L. N. Kanal, J.F. Lemmer, *Uncertainty in artificial intelligence,* North-Holland, New York, 1988.

[Lamdan88] Y. Lamdan and H. J. Wolfson - "Geometric hashing: A General and Efficient Model-Based Recognition Scheme". Proc. of Second ICCV, 1988, 238-289.

[Latombe91] J.C. Latombe, "Robot motion planning", *Kluwer Academic publishers,* 1991

[Lee90] C.C. Lee, *Fuzzy logic in control systems: fuzzy logic controller;* (Part I and II) IEEE Trans. on Syst., Man and Cybernetics, Vol. 20, No.2, pp. 404-435, 1990.

[Lieury04] Lieury, A. (2004). *Psychologie cognitive* (4ème édition). Paris, Dunod.

[Lowe87] D. G. Lowe - "Three-dimensional object recognition from single two dimensional images". Artificial Intelligence, 31(3), 1987, 355-395.

[Mills90] D. L. Mills. On the Accuracy and Stablility of Clocks Synchronized by the Network Time Protocol in the Internet System. *SIGCOMM Computer Communication Review,* 20(1):65-67, 1990.

[Moon00] S. B. Moon. Measurement and Analysis of End-to-end Delay and Loss in the Internet, 2000. Diploma thesis.

[Peruch99] Péruch, P., & Mestre, D. (1999). Between desktop and head immersion : Functional visual field during vehicle control and navigation in virtual environments. Presence, 8, 54-64.

[Piaget36] Piaget, J. (1936). *La Naissance de l'Intelligence chez l'Enfant.* Paris, Lausanne : Delachaux et Niestlé. Translated in 1952.

[Piaget52] [PIA52] Piaget, J. (1952). *The Origins of Intellegence in Children.* N.Y.:The Norton Library, WW Norton & Co, Inc. Translation of [Piaget36].

[Raftery95] Raftery A. E. Bayesien Model Selection in Social research. Sociological Methodology, Vol. 25 (1995), 111-163.

[Roulin06] Roulin, J.-L. (2006). *Psychologie cognitive* (2ème édition). Bréal.

[Zadeh65] L.A. Zadeh, *Fuzzy sets;* Information and Control, 8, pp338-353, 1965.